Architectural Gardens

Architectural Gardens

INSIDE THE LANDSCAPES OF LUCAS & LUCAS

Thad Orr and Mike Lucas

PHOTOGRAPHY BY CAITLIN ATKINSON AND MARION BRENNER

PRINCETON ARCHITECTURAL PRESS · NEW YORK

Contents

Preface

THAD ORR

It's hard not to be drawn into a Mike Lucas landscape. Five years ago, I laid my eyes on a photograph of a vignette from one of the landscapes Lucas & Lucas designed. I was immediately enthralled. The lush, elegant, and cozy courtyard was beautiful on its own, but the built-in landscape windows were unlike anything I'd ever seen.

As I explored more Lucas & Lucas landscapes, I discovered other inventive features: planted steps, cascading concrete waterfalls, boulder seats. The diversity of styles also struck me. The firm's designs don't fit into a particular genre but, instead, adapt to a home's architecture and its surrounding land. There are modern and mission-style houses, old-world estates, and traditional dwellings with wraparound porches. Though Lucas applies common principles, he avoids prescriptive formulas. Each landscape is distinct.

Lucas & Lucas is a small, service-oriented firm, led by Mike Lucas and Jennifer Lucas, who are a husband-and-wife team. Founded in 2005, the practice is an incarnation of their balanced lifestyle and core values, rooted in the community and landscape of Northern California. By keeping their practice small, they maintain a close relationship to their process, creations, and clients.

The story behind each design is as engaging as the landscape itself—how they came to be, what decisions were made along the way (and why), and how the gardens evolved through the design, construction, and planting phases. I felt that his ideas needed to be shared. Together, Mike and I selected the ten projects in this book to offer inspiration and information to the readers seeking to create architectural gardens of their own. Videos about each project are available on the Lucas & Lucas website (www.lucas-lucas.com) and offer an even closer look at Mike's designs, whether it's the way the shadows move within the garden or how light hits a particular plant or water feature. These videos also provide a deeper understanding of the gardens and the thought process behind their creation.

Each Lucas & Lucas design inspires insights for home gardeners, professional designers, architects, and anyone interested in gardens. There are ideas for specific features (landscape windows, planted steps), ways to capture the elements (water features, grasses that catch the breeze and sunlight, oaks that cast shadows), and conceptual ideas about the design process (exploring the land before designing, making the best use of trees, developing stunning views). Take notes and have fun exploring Mike's creations.

Introduction

MIKE LUCAS

Architectural gardens create a union between the land and home, the indoors and outdoors, the natural and constructed. The goal of any designer should be to create a landscape that balances these elements in a beautiful and useful way.

My views on the connection between architecture and the garden began while I was preparing for a master's degree in architecture. I had stumbled into the landscape architecture department during an ecological building conference at the University of Oregon. I remember peeking my head into a studio and seeing all sorts of colors on the wall, illustrative drawings of stream trails, gardens, and urban landscapes. I was immediately drawn to them. By the time I returned home from the conference, I had decided to pursue a degree in landscape architecture instead.

Before this experience, I had no idea that the discipline existed (many people don't). As I began the landscape architecture program, the study of plants, nature, outdoor environments, drawing, sketching, and design captivated me and continue to do so to this day. On some level, I felt that I'd been preparing for a career in landscape architecture all my life (unbeknownst to me). I have always been a lover of nature, enjoying hikes, surfing, and all things outdoors. I'm also a homebody. My affinity for architecture grew out of a love for the arts, sculpture, form, drawing, and ultimately, a love of home. Landscape architecture at the residential level married all these realms.

Since my graduate school days, I've seen the two fields—architecture and garden design—as inextricably linked. All gardens extend from the home. When I begin work on a property, I start with the architecture of the home and work my way toward the outer reaches of the property. Whether the home has already been built or I'm working with the architect during site planning, the purpose is the same: to create an environment in which owners and their guests can move comfortably and easily between indoor and outdoor spaces.

Working in Northern California has allowed us unparalleled opportunities to work with a wide variety of environments. Discovering what a property has to offer is often a process of exploration—of the land itself, how the architecture fits into the land, and the owners' desires. Rarely is the design direction evident at the start of a project. It's usually revealed through numerous site visits and meetings with the owner and architects in an effort to see how various design options work, how the owners respond to the choices, experimenting with materials, and making informed revisions at each stage of development, as layers of information unfold through our iterative process.

A Mentor's Influence

Being open to this type of exploration and collaborative design is what leads to the creation of a special landscape—remaining open to the possibilities and building upon a solution as it gains momentum and comes into being. I first came to fully understand this approach to landscape design when I met Gary Ratway, a landscape designer based in Northern California well-known for creating thoughtful gardens that respond to the land, the house, horticulture, and the homeowner.

I'll never forget the first day I jumped in Gary's old van to visit three projects that were in various stages of construction. I had never experienced landscapes the way he had designed them—a visceral feeling overtook me as the spaces within each garden unfolded around the homes we saw. The gardens led me from space to space, guiding me to the best spots on the property as specific views were revealed. Everything—stairways, walls, plants and trees, walkways—was perfectly balanced and befit the home's architecture. Every step through each garden felt right. I absorbed every view, scent, and feeling from those first few gardens we visited.

Over the decade we worked together on various projects while I developed my own firm, I gained a deeper understanding of the practice of landscape architecture and how it is done in the field. I learned how to interact with clients and architects and how to suss out a client's desires for their property. I came to understand how to observe the land in its raw form and how to guide a design through construction. I was able to fully appreciate how plants can be used to create distinct spaces and to soften architectural features or reinforce them, highlight a special view, or blur the lines between architecture and garden.

Many of the design principles I lean on in my practice today were gained during these early years working with Gary. I learned from him through osmosis, observation, and participation in his creation of structured yet informal gardens that are a joy to explore. His gardens also set an example of how to fill spaces with exuberant plantings that create a sense of movement and liveliness. I can see many of the lessons I learned early on from working with Gary reflected in the projects featured in this book.

Thinking about my time with Gary and the work I've done since, I appreciate how his gardens never fit into one style of architecture. The process and evolution of each of his gardens is a unique story and solution unto itself. He doesn't design traditional gardens, modern styles, or Italian villas. At Lucas & Lucas, we've made an effort to do the same. As you'll see in this book, our projects don't fit into a particular style or approach to planting. Each garden fits the land it's on, the architecture it relates to, and the owners who live within it.

Exploring Possibilities

No two projects are the same. We do not have a prescriptive approach or a signature style. The results are always as varied as the owners, architects, and sites. Of course, we guide, consult, and sometimes prod owners and architects toward a particular design solution when there's a reason. Generally, though, the approach is collaborative. It's a process of finding many solutions that fit together into a cohesive whole.

When projects begin, solutions are rarely obvious or readily available. They are revealed and teased out through trial and error—through design sketch after design sketch, site walk after site walk, discussion after discussion. Ideas and approaches are adjusted and refined until something clicks with the place, the people, and the architecture. I know we've found a solid solution when the design team, which often includes the owners and the architect, sees the design and says, "Yep, that's it. Love it."

Like solving a good puzzle, the challenge of creating a plan is highly enjoyable and invigorating, whether sitting at the drafting table thinking about what's possible, visiting the site to feel and explore different spaces in an effort to see whether ideas will truly work, or talking through different design solutions with the team. I often have initial impressions and ideas when I visit a site, which I pursue in a design sketch. Some of those ideas stick and become the driving force of the design, while other ideas are dropped as less obvious solutions emerge.

Often, major decisions stem from the design of the home and the functional spaces inside. From my perspective, the home is always the center of the garden. The interior of a home informs what happens outside: I imagine how movement will flow between spaces and what functions will support one another and seamlessly blend to result in an experience of living in the space that is effortless and feels right. Practical considerations are the core of the design: making the space comfortable and functional with shade and warmth, providing access to the house, designing for easy maintenance, and selecting plants and materials for long-term durability are all vital considerations when developing a garden.

It's All in the Details

Once I'm sure that practical considerations have been resolved, I can focus on making it beautiful (though a certain level of beauty arises out of the pragmatic solutions themselves). This is often achieved by making natural elements more visible in the garden, whether adding a seating area under an oak tree that captures the light perfectly, surrounding a garden with grasses that sway in the breeze, creating a reflection of the sky in a vanishing-edge pool, or showcasing gorgeous shadows from trees on a pool deck. Such elements artfully bring nature into the built environment and help connect the home and its living spaces to the land.

Planting is the softness, the wisps of texture and life, that comes in at the end of the installation. Instead of developing strict planting plans before a garden is built, I like to let them evolve as the bones of the garden are being built. During construction, as the plans are coming to life on the ground, I get a better feel for spacing, soils, shade, and sun, as well as other variables unknown during site analysis and design.

Adding customized architectural details that fit with the architecture of the home also helps beautify a landscape and connect it to the home. A water feature designed for a home's entryway is more of a sculpture than it is a fountain and catches the afternoon light I noticed day after day as I returned to the site. Windows cut into a courtyard wall showcase stunning plants like a classical landscape painting come to life. You can even beautify steps by adding plants to enjoy along the way.

Wonderfully, as we are building, we can edit and change the gardens. Some of the best ideas emerge during the construction process or come from a collaboration with artists. I've worked with stonemason Edwin Hamilton for many years, and he brings his creativity and expertise to each project. Clever techniques, such as slicing boulders to create slab steps or building stone walls with extremely tight seams, have helped make many of these projects shine.

Refining the Design During Construction

As much as we discuss and research during the planning stages, there's nothing like building a garden to unsettle the nerves—mine and those of the owners and the architects. Building a landscape can be a long and arduous process that can take many months or even years. At times, it's all about having faith through construction and reinforcing the ideas that were formed during design. I have to trust in our decisions too! When the land is being torn up and the pipes are being placed into the ground, it can be hard to visualize the final product. Eventually, trenches are backfilled, hardscape is laid out, and spaces start to take shape. These built elements are the true bones of the architecture in the landscape, and they must be precise.

When the landscape is finally being built, owners can better understand what we have been discussing and respond to what's being installed. Each layer added brings energy and depth to the garden architecture, and the place slowly comes into being as final decisions are made, materials installed, and plants placed into the ground.

The ten gardens profiled in this book represent what is possible when home and garden are conceived as one. They come to life through a push and pull that occurs during the design and construction process. This is the work I love the most: the refining of ideas, the balancing of various elements, the transformation from bare soil or a dated landscape into a living sculpture that people can move through enjoyably and inhabit with friends and family.

CHAPTER ONE

Eleven Design Lessons

Thousands of decisions, both large and small, go into the creation of a landscape that is comfortable and beautiful. The goal is to create a refined design that feels like it belongs on the land, fits with the architecture, and is loved by its owners. “To guide the process, I keep a few overarching concepts in mind as I explore a garden and think about what’s possible,” says Mike Lucas. “These concepts help create consistency and continuity in the final design.” He shares them here.

1. Let the Site Speak

Before designing a garden, Lucas explores and observes the land closely to identify the most important existing elements. His goal is to make those moments visible and accessible to the homeowners. Along the way, he adds and edits to enhance the beauty of the property while choosing materials and plantings that look as though they were always part of the land. "I want to highlight the natural beauty of each site," notes Lucas. "Sometimes this means I do very little on a project, focusing more on how people move around a site comfortably and intuitively. Other times, this means constructing views and visually appealing architectural features. I like to let the land guide me."

2. Design for the Best Views

Design seating areas, entertainment spaces, lookouts, or even walkways that capture the best views. Removing vegetation from one area may reveal a grand vista in another area. "I enjoy spending time getting to know every inch of a garden. While touring the land, I imagine views with various plants, how a hillside might look with swaying grasses lit up by the sun or the shadows cast on paving under gorgeous oaks. Visualizing what's possible is a vital step in the design process and for discovering the best views—whether they are grand vistas or intimate spaces within a garden," says Lucas.

3. Think Like a Sculptor

Think about what should be removed and what should be kept before deciding what to construct. "I tend to be methodical and spiritual in my approach to developing a new plan," Lucas says. "I find that if you move too fast you can blow right by the character of a landscape." For example, on many of his projects, specimen oak trees are preserved, carefully pruned, and incorporated as features of the new landscape. Rather than embarking on extensive grading, consider carefully carving retaining walls into the hillside around existing trees, if necessary, to keep their roots undisturbed (it is always a good idea to consult with an arborist in these sensitive situations).

4. Entice with Scenic Paths

Walkways are not just about getting from point A to point B—they also provide an experience. Like the curvy branches and trunks of an oak tree, paths can be free-flowing and organic. They provide access to special areas of the property but also create enjoyable journeys: along the paths, grasses move in the breeze. Beautiful stone boulders emerge from the land. Flowers bloom in vibrant colors. Light streams through the trees. Shadows dance on the ground. In other instances, paths may be straight but envelop visitors in exuberant plantings, overarching rose-covered arbors, and calming fragrances. "Walkways often greet people when they arrive and create a lasting impression when they leave. They should create a memorable emotional response."

5. Celebrate Natural Habitats

Respect the site's flora and fauna. Preserve existing ecosystems while celebrating their beauty.

On one site, vernal pools fill up each spring, attracting frogs, birds, and other wildlife. Rather than alter this stunning feature, Lucas created an inviting pathway that sensitively moves through it. On other projects, Lucas has observed that existing oak trees support abundant local wildlife, including native species of birds and butterflies. Rather than replacing them or possibly damaging their roots during construction, Lucas works to preserve and enhance their health and beauty.

6. Blur Boundaries

Remove fences and other hard boundaries and replace with plants, water, or natural stone to allow views to transition to the horizon seamlessly. When fences, borders, or gates are necessary, Lucas uses plants and shrubs to hide visually distracting elements. Tall, tawny grasses blend smoothly with annual grasses and block smaller unsightly features. Larger shrubs, such as coffeeberry, bay laurel, or coast silk tassel, enhance the visual appeal of property lines. "I want people to have to look really closely to know where their property ends," says Lucas.

7. Limit the Hardscape Palette

Ensure that the colors of new interventions integrate with the palette of the materials that are found on the site. Colors need to coordinate with those of the trees, grasses, soil, and stone around the property. New materials will look out of place if they don't remain true to the feel of the land itself. For example, on projects where oaks, native grasses, and boulders are prevalent, Lucas opts for a neutral palette. Gray board-formed concrete retaining walls echo the gray bark of the oaks. A wood pool deck and seat wall merge with brown grasses. A dark gray basalt walkway and steps coordinate with the colors in the native soil and boulders.

8. Soften Edges

Anywhere architectural hardscape or walls form hard lines, add plants to soften the edges. For example, Lucas might plant cascading creeping raspberry over the tops of retaining walls. He builds paths in a saw-toothed pattern or breaks them up to minimize the number of straight lines, leaving gaps between paving or along their edges to make room for plantings that soften the lines.

9. Repurpose Unused Materials

As a project nears completion, imagine new uses for leftover materials from paving, walls, or other architectural features. Instead of discarding or recycling the materials, Lucas repurposes them for smaller architectural features, such as raised beds, fountains, or seating walls. "Reusing these leftovers can turn a utilitarian area of a property, such as a kitchen garden, into an artistic space," says Lucas.

10. First Impressions Are Vital

The arrival experience sets the stage and welcomes guests. Add rows of trees or create a parking area where cars aren't in full view from the house or landscape. Move a driveway or regrade the front of a property. "Imagine driving along a meandering driveway surrounded by vineyards. Then imagine driving along a barbed wire fence. There's one clear approach that sets the proper tone for a landscape," says Lucas. After driving through a vineyard, walking up a path with the crunch of gravel underfoot, the experience of approaching the home among old olive trees is one that leaves a lasting impression.

11. Entice People to Explore

Steps, walkways, sculptures, water features, and glimpses of a structure entice people to explore. A distant roofline or set of steps creates wonder and curiosity. People gravitate toward water or water views. Including some of these attractions draws people into the landscape.

CHAPTER TWO

Knolltop Retreat

Overgrown trees and shrubs once obscured views to an oak-covered meadow and the hills beyond. Lucas placed a pool and pool house on the edge of a hillside to maximize views.

When Mike Lucas first encountered this hilltop estate just outside Santa Rosa, California, gangly, twenty-foot-tall hedges blocked views of the surrounding hills. Rhododendrons and other exotic shrubs were planted incongruously under native oaks, and redwood and pine trees grew throughout the property; irrigation ran frequently to keep these plants looking lush. A dilapidated tennis court occupied the middle of a grassy meadow.

"The existing landscape felt artificial and inappropriate for the location," Lucas says. To understand the land, Lucas visited the property week after week for more than six months while a remodel of the house was in its early stages. "When I start a project, I need to live with it and feel it: I explore every inch of the site," says Lucas. "I sit in locations that have grand vistas. I walk where paths could be made. I explore areas that are overgrown to see what is being overlooked. I look at the colors of the bark, feel the winds, watch how light cascades through the trees." These observations help him discover what belongs on the land and what needs to be edited, driving major decisions.

Lucas started by tagging specific oak trees to be preserved. Next, he removed the overgrown water-loving shrubs and perennials under the oaks. This opened up the oak canopy, revealing the sculptural forms of the majestic branches of the trees. Hedges and redwoods were removed from the property lines to create sweeping views and to ensure that the land would blend with the surrounding oak woodland.

This editing process created space in the landscape for paths, fountains, and seating and entertainment areas. Lucas paid special attention to the design of the new features, keeping them simple and appropriate for the site. He chose hardscape colors that blend with the earth tones of the nearby annual grasses, bark, and native soil. Paths follow the contours of the land. Walls are tucked into hillsides.

The homeowners brought Lucas in to design the landscape after construction on their home renovations had commenced. Knowing that some of the decisions about the orientation of the house and the building materials had already been finalized, Lucas worked to integrate the home's materials and colors into the landscape.

Lucas wanted to extend living spaces farther into the property, to allow the homeowners to enjoy beautiful views. The placement of the pool was a key decision. The original plan had been to install the pool immediately to the back of the house; instead, it was placed in a large meadow beyond the old tennis court, which was removed. Its tucked-in position protects the area from afternoon winds and provides western views otherwise unavailable from the house. It's now a destination spot with gorgeous vistas all around. "It's the most calming place to spend the afternoon—enjoying the day fade into evening as the hills turn golden," says Lucas.

The distant location of the pool area also created opportunities to build pathways that lead from the house to the pool and back. Two main paths meander under oaks and through understory plantings that include slender veldt grass and moor grass, as well as perennials, such as Douglas iris, kniphofia, aster, yarrow, and salvia for seasonal color.

"The paths are peaceful. They're great places to watch the wind blowing through the trees, the shadows of the oak leaves, and the light cascading through the branches. It's calm and relaxing," says Lucas. Secondary paths turn off to other significant destinations: one path leads to a trampoline at the top of a knoll, another splits off to a bocce court on the side of the property, to a spot Lucas discovered during his early explorations. The bocce court has a uniquely stunning view of the property and hills beyond.

Another major decision involved the parking area. Originally, cars entered and parked at a courtyard. When the homeowners walked out the front door, those cars dominated the space, so screening them from view would make a significant visual impact. Lucas moved the parking area away from the house, placing a few small board-formed concrete walls to hold back the hillside and to screen the cars. Grasses, shrubs, and trees also help obscure the parking area. "Now you can exit the front door, look out a window, sit on a bench in the courtyard, or walk down a path without ever knowing cars are there," says Lucas. "It's as if everyone on the property hiked there. You do see a few walls, but the focus is on plants, trees, boulders, and the sky."

Moving the parking area out of the courtyard created another opportunity for gorgeous paths. These paths, leading from the arrival areas to the front door,

The light in the grassy meadow beyond the pool cascades down a gentle slope. Low-growing grasses, such as Greenlee moor grass and slender veldt grass, blend with native grasses, blurring boundaries between the garden and the wildlands. Solar panels are strategically tucked into the meadow behind a planted earth berm.

The pathway from the driveway to the front door is fully paved and wide enough for two people to walk on comfortably, side by side.

with moments of respite along the way, create spaces for a welcoming arrival or a grateful farewell. Outside the front door, an L-shaped seat wall creates the perfect spot for hellos and goodbyes.

A rectangular fountain near the parking area is the first landscape element that guests encounter. The soothing sound of water trickling from the fountain is audible along the paths. "I didn't want a large, loud, ornate fountain," says Lucas. "But I did want it to be heard, as if to say, 'Come. This way to water.'" In tune with the fountain's setting, water pours from a boulder and spills into a simple concrete basin that's tucked into the hillside, reminiscent of a natural spring. "There's nothing overly complicated here," says Lucas. "With every decision we wanted to maintain the integrity of the land and not distract from it."

The boulders, one of which was carved for the fountain, are a critical part of creating a natural ambiance. Lucas included additional boulders near the fountain on either side of the path, and throughout the property, some of which do double duty as retaining elements. He placed them strategically to make it seem as though they had always been on the property. "In nature, you rarely see boulders all by themselves. They are typically found in groups," he says. The native basalt variety he chose has a variegated finish that coordinates with the mottled bark of the oak trees.

He also chose basalt paving for a majority of the walkways and patios throughout the landscape. "I was careful to keep the material palette simple and understated," says Lucas. The dark gray color also coordinates with the house colors, including the gray roof and tan siding. In total, only four hardscape materials are used on the property—basalt for main pathways and patios, wood for the pool deck, gravel for a few smaller paths, and board-formed concrete for walls.

This simplicity helps elevate the wild, naturalistic features of the site that Lucas experienced when he first visited: the views, the trees, the plants, the sky, the light. "When designing gardens, I'm helping people discover and appreciate their property," he says. "I make what's most special about it accessible so they can build a better relationship with the land they live on." ☼

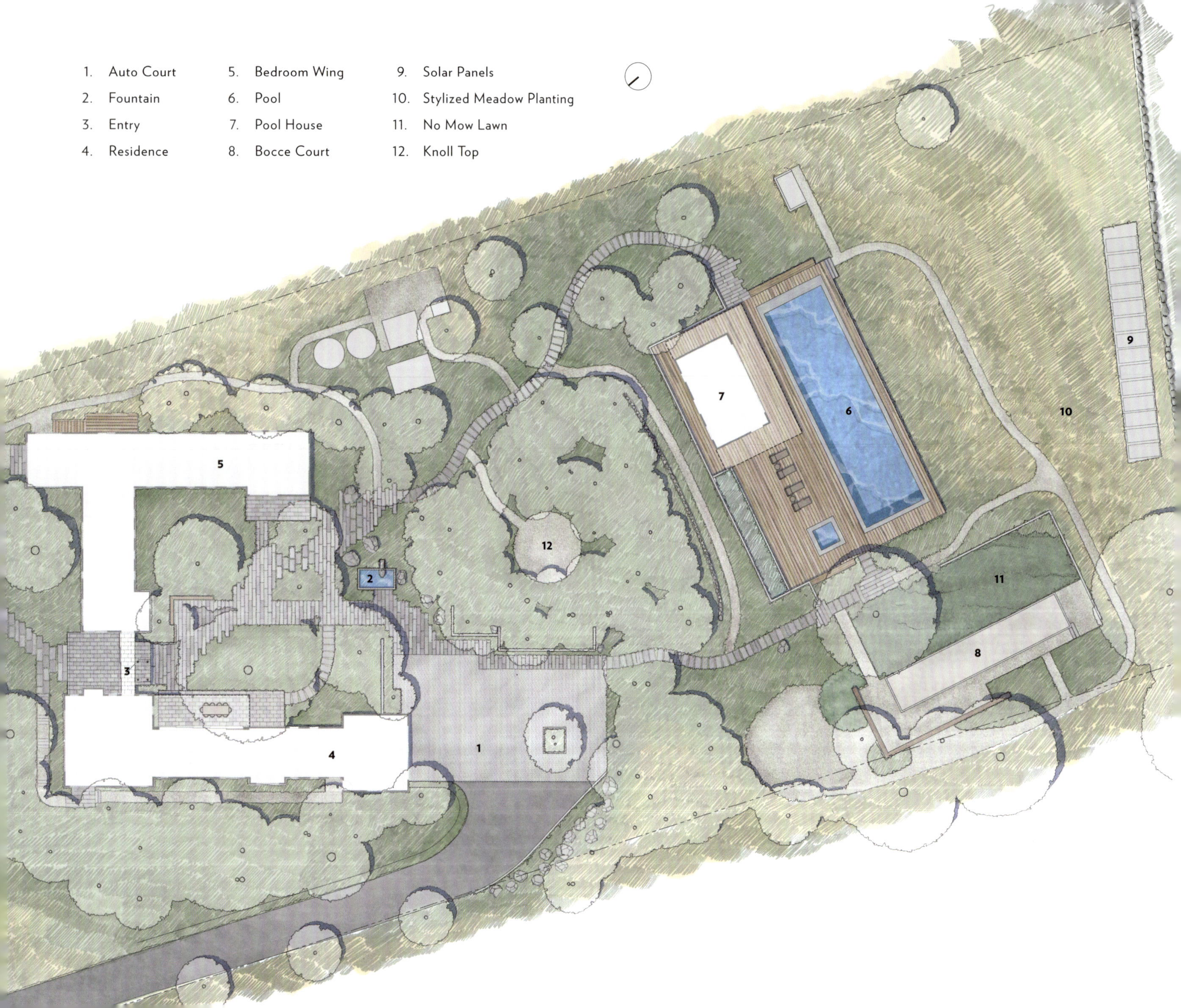
1. Auto Court
2. Fountain
3. Entry
4. Residence
5. Bedroom Wing
6. Pool
7. Pool House
8. Bocce Court
9. Solar Panels
10. Stylized Meadow Planting
11. No Mow Lawn
12. Knoll Top

ADVICE FROM LUCAS

Fountains That Fit Right In

Fountains should fit the scenery. This fountain is by no means grand, but it feels natural tucked among groups of boulders. The boulder over the edge of the concrete basin was carved by hand to provide just the right water flow. Acoustics in the garden are a key consideration: the sound of a fountain should be audible but not overpowering. Plants tucked in around the fountain base help to integrate it with the land; here, I used juncus, Greenlee moor grass, *Heuchera maxima*, and asters.

ADVICE FROM LUCAS

Preserving Oaks

Make specimen oaks the focal points of the garden. If you're lucky enough to have mature oaks, carefully preserve them during landscape projects. Not only do they provide a sense of instant maturity, but they also support a wide range of wildlife, and their sculptural beauty is nearly impossible to replicate. I like to run paths under oak canopies where filtered light streams through. Siting a pool between two large oaks can give it a sense of enclosure and safety.

Though it's often best not to cut back oaks, it may be necessary where branches are damaged or diseased. Always consult an arborist to ensure that any branch trimming or work around the roots won't damage the trees.

We took out water-loving plants around some of the oaks (because excessive water can slowly kill oaks) and replaced them with low-water plants, such as ornamental grasses, asters, salvias, and coffeeberries. We cut the water back slowly throughout the first few years; the newer plants appreciated additional water as they became established.

Autumn moor grass softens the edges of offset basalt paving. Asters grow behind grasses, providing color in later summer and fall. Shallow steps broken up by landings accommodate the changes in the grade gently, preserving the feeling of a stroll through the woods.

LEFT The earth tones of the ipe-wood pool deck, stone steps, and board-formed concrete walls connect with the colors of the native landscape. These rugged materials age beautifully with minimal maintenance.

RIGHT A small retaining wall holds back the slope at the end of the lap pool. It also provides a spot to relax after a swim or to enjoy the golden sunset.

LEFT Basalt pavers form a gently sloping path at the southern side of the pool. Small gaps in the paving form a rhythmic pattern down this roughly 150-foot walkway, which meanders through a grove of oaks. Along the way, paths jut off toward the house or to play features, such as a trampoline on the top of a knoll.

BELOW LEFT Grasses work well to soften architectural lines because they tumble over edges and bring movement to the landscape. Here, Lucas planted *Pennisetum spathiolatum*.

BELOW RIGHT A path through the meadow beyond the pool encourages exploration. Yellow 'Moonshine' yarrow, *Verbena bonariensis*, salvia, and agastache add color to a sea of tawny *Pennisetum spathiolatum*.

LEFT Greenlee moor grass, *Heuchera maxima*, and other low-water perennials, such as aster, salvia, and kniphofia, replaced scrubby, water-thirsty plants under the oaks. This substitution opened the view to the back of the property, simplified maintenance, and created space for curvilinear paths that reflect the canopy of twisted oak branches and trunks above.

LOWER LEFT A meadow beyond the pool obscures solar panels lower on the hillside. Before the landscape redesign, this view was obscured by overgrown shrubs. Now this is a destination within the garden for exercise and entertainment.

RIGHT Narrow paths leading from the house to the bocce court, meadow, and pool nearly disappear into the grasses. This effect is reversed when returning to the house: the paths widen and lead the way to the entry.

BELOW LEFT The arrival and parking area creates a welcoming mood with wide paths and low walls. These board-formed concrete walls keep parked cars out of sight and provide transition points between the built landscape and the land.

BELOW RIGHT The basalt steps to the pool are staggered, creating spots to tuck in plants.

RIGHT To tie in the view of the adjacent field, Lucas planted a few varieties of tan flowering grasses.

LEFT The view from the home's living area is tucked away behind a preserved oak tree, as well as behind coffeeberry shrubs and ornamental grasses. Though set in the entry courtyard, this space feels private and peaceful, with views of the oak woodland in the distance.

RIGHT Removing redwoods, rhododendrons, and other scrubby plants and carefully trimming back the oaks allowed sunlight to stream through the canopy. The wide basalt path leads from the front door through numerous Greenlee moor grasses. Since this variety of grass is an evergreen that doesn't need to be cut back in winter, this area stays verdant year round.

CHAPTER THREE

Private Sanctuary

A large rectangular garden window forms a living painting in the courtyard walls. A Weber agave is surrounded by colorful orange flowering knifophia and purple flowering 'Mesa Azure' salvia.

When Mike Lucas was commissioned to develop the site plan and landscape for this home in Sonoma, California, the property, where an old pear orchard had once stood, was overgrown and dilapidated. Barbed wire fences were strewn throughout the site. Coyote brush, poison oak, and tall grasses choked off access to much of the six-acre property's ephemeral streams, majestic oak trees, towering eucalyptus groves, and seasonal ponds, along with the wildlife they attracted. "It had that abandoned ghost town feel, but I could tell that there were beautiful features hidden within," says Lucas.

Though his clients had already enlisted the services of an architect, the design process was in its infancy when Lucas came on board, and he was able to collaborate with the architect and the homeowners before a plan for the house was developed. He incorporated all of the owners' wishes—mainly a relaxing space to dine and entertain outdoors—and addressed the site's major challenges, including brutal winds, a nearby road, and well water with high levels of boron.

Lucas's design team found inspiration in the local Mission San Francisco Solano, where courtyard walls provide refuge from the elements. The homeowners and designers also incorporated aspects of the mission's spare aesthetic into the home's design. Architectural elements, such as the tower, courtyard walls, and shade structure, punctuate the otherwise flat site. "These features create focal points for outdoor entryways and spaces to entertain," says Lucas. "Trees were the only vertical visual elements on the otherwise flat site."

Among those mature trees are large groves of drought- and boron-tolerant Australian eucalyptus, which serve as a cornerstone of the planting plan outside the courtyard walls. "The trees bring scale and age to the land," says Lucas. "They have a wonderful ambience with their beautiful shapes and the way they catch the light." The driveway gate is tucked into a stand of eucalyptus trees. "The trees set the tone for what's to come," says Lucas. Along the approach to the mission-inspired home, most of the plants are also Australian and boron tolerant: 'Orange Cross' kangaroo paw, lomandra, acacia, peppermint tree, bottlebrush, and coast rosemary.

Other selections were planted in clusters or as individual specimens strategically placed to obscure the road. They also hide water tanks that store potable water, which is used in the house and for boron-sensitive plantings in the courtyard.

Outside the courtyard walls, Lucas selected plants for their resilience and durability. Inside the courtyard, he created a green oasis. "My favorite moment in this garden is opening the courtyard door," says Lucas. "When you step in from that arid landscape, you're greeted by the sound of water spilling from a reflection pool. Plantings are lusher than those outside the walls and in richer shades of green. It's quiet and tranquil."

In addition to a flowing fountain and soft greenery, the courtyard also has a firepit that provides warmth, a pergola that offers shade, and strategically placed plants that give off fragrance. Simple gravel paths lead through verdant stands of agapanthus. Sweet-smelling groups of jasmine line walkways. Mock orange adds a citrus scent. A 'Mission' fig tucked into the corner of the courtyard produces ample fruit for the summer harvest. "The environment inside the courtyard delights the senses, and its simplicity is visually calming," says Lucas.

The development of the courtyard started with its walls. Lucas opted for eight-foot-tall walls, rather than the six-foot standard, to block the wind, to minimize much of the noise from the nearby road, and to obscure less-than-ideal views. Since the property lacks views typically available on hillside properties, Lucas decided to make their own views. From inside the courtyard, the view over the walls reveals towering eucalyptus trees and distant hills—with no road or neighbors in sight. Lucas often uses plants to accentuate views. In the courtyard, he grew Boston ivy up the walls—layers of green—making it easier to focus on the distant view of trees and hills beyond.

To reduce a potential sense of claustrophobia in the courtyard, he cut two roughly eight foot by four foot windows into the large walls. These windows frame stunning views of the landscape beyond—highlighting hills and trees, flowers in bloom, or plants with particularly gorgeous forms, such as agaves. The windows were fitted with rolling shutters so that the owners can close them when desired.

The courtyard is divided into open-air rooms—some rooms are elevated to the level of the house (about two feet above the ground), while others are at grade. Elevated patios abut the L-shaped house, from which the homeowners can easily access the home's

Drought- and boron-tolerant plants fill planting beds on the outside of the courtyard walls that surround this mission-inspired home. Orange flowers of Australian native 'Orange Cross' kangaroo paw stand out against the cream-colored courtyard walls, creating a dramatic entrance.

Trees and a metal-and-wood shade structure shelter the outdoor dining space. A path lined with blue-flowering agapanthus leads toward the dining area. A fig tree is tucked into the corner of the courtyard.

kitchen and two of the bedrooms. These patios are raised a few feet above the courtyard, creating slight but impactful topographical changes within the courtyard. Intimate outdoor spaces—such as the dining area that sits under a pergola and the firepit with built-in seating—are tucked into the far corner of the courtyard, which is the main gathering spot.

Lucas uses plants and landscape elements to draw people out of the house and into the landscape. Visible through a breezeway toward the back of the house is a linear walkway lined with ornamental pear trees and jasmine ground cover, which leads to a large bowl-shaped fountain. "It was important to get people out to the 'back forty,' as the owners call it," says Lucas. "It's so magical back there, where the old pear orchard once stood, with the hawks flying overhead, the sun streaming through the oaks, quail scurrying across the ground, and seasonal streams, ponds, and frogs," says Lucas.

Beyond the allée of pear trees and fountain is a twelve-foot-wide, mulch-covered pathway that leads through oaks and groups of eucalyptus trees. "Nearly every weekend throughout the construction process the owners were out there clearing the path by hand," says Lucas. It's now a walking trail the owners use regularly (the wood chips were generated from their clearing efforts). They can observe wildlife while they stroll, and there are even a few places to sit and relax. There's an enormous bench made from a fallen eucalyptus tree and large stone slabs that surround a firepit.

Lucas believes that the best designs come from thoroughly exploring the land as a team—the homeowner, architect, and landscape architect. He enjoys figuring out the important concepts together: How will the home fit on the property? Where will those special moments take place? What is the most exhilarating way to drive onto the property or approach the front door? What are the best views? "You can't visit a property once, sit down at the desk, and come up with this type of design," says Lucas. "The landscape has to evolve with the building. That's the true magic."

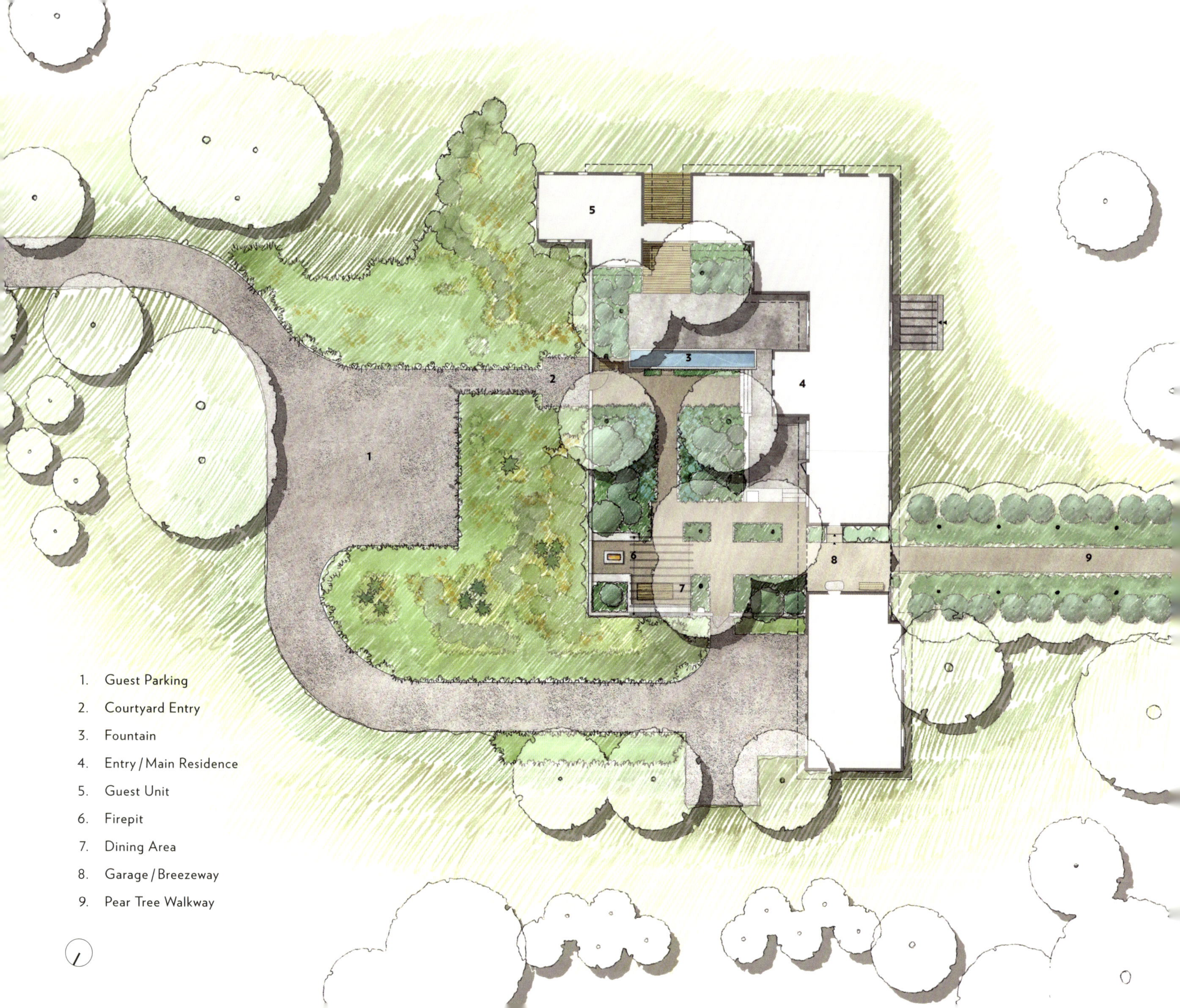

1. Guest Parking
2. Courtyard Entry
3. Fountain
4. Entry / Main Residence
5. Guest Unit
6. Firepit
7. Dining Area
8. Garage / Breezeway
9. Pear Tree Walkway

ADVICE FROM LUCAS

Creating Landscape Windows

Use windows and doors through walls to frame views. Due to regular afternoon winds and the noise of a nearby road, the client desired an enclosed space for dining, mingling, and cooking outdoors. To maintain a visual connection between the courtyard space and the landscape, windows and doors were punched through the walls to frame views; old-fashioned rolling shutters are deployed when the winds get too intense. The doors align with walkways that cross the courtyard and extend views into the greater landscape.

To allow for large windows, the wall height was set at eight feet, which provided enough space around the window for structural support. During the design process, the owners, as well as their friends and family, were skeptical about the height. We had regular discussions during which I explained my intent of avoiding the experience of a six-foot fence, which would be too easy to see over and wouldn't block wind or noise well.

We mocked up the size of the wall with string to provide a visual. After some discussion, the owners, although still unsure, decided to trust me and go with the eight-foot walls and large windows. They have been a hit with the owners and guests ever since.

ADVICE FROM LUCAS

Delighting the Senses

Fill the space with fragrance, which, like the sound of water, can imbue a place with a sense of serenity. I try to incorporate fragrant plants into most gardens, and this courtyard was no exception. Fragrant jasmine, *Osmanthus fragrans*, *Sarcococca ruscifolia*, mock orange, citrus, and roses fill the courtyard—along paths and in the dining area—with wonderful scents throughout the seasons.

ADVICE FROM LUCAS

Thinking About Wall Coverings

Soften walls with lush vines to connect structures to the garden. The interior walls of the courtyard are covered in Boston ivy to minimize the visual impact of their eight-foot height while ensuring that they blend into the landscape. Conversely, the courtyard's exterior walls are covered with white plaster to match the exterior of the house. This creates a juxtaposition between the tough exterior of the home with the soft interior of the courtyard.

LEFT Vibrant blooms of 'Orange Cross' kangaroo paw and the grasslike foliage of *Lomandra longifolia* 'Breeze' form a dry meadow that's fitting for the climate and appropriate with the Australian plant palette outside the courtyard.

RIGHT Weber agave is planted among grasslike lomandra and kangaroo paw.

LEFT Shade, fragrance, a gentle breeze, food, the sound of the water—all combine to create a feeling of comfort.

RIGHT The long water feature acts as a reflection pool, bringing the trees, sky, and sunlight into the courtyard. Creating connections between the courtyard and the garden outside its walls was a top design priority.

LEFT A small number of plant varieties create a tranquil and verdant courtyard interior. A long trough fountain spills into a hidden basin under the gravel. Boston ivy covers the walls, creating layers of greenery.

RIGHT The view through one of the garden windows shows a repurposed gravel crusher that Lucas found in the boneyard of a local quarry and employed as a fountain, encouraging the exploration of the back of the property. Weber agave, lomandra, 'Little Ollie' dwarf olive, and tall verbena fill the planting bed in front of the courtyard wall.

CHAPTER FOUR

Hillside Modern

Built into a hillside, this property overlooking Sonoma Valley has a series of stairs, landings, and seating areas that turn the slope into a landscape with many useful spaces. Seat walls, a game area, a bocce court, and driveway are built into terraces created by a network of retaining walls.

"When I arrived on the property, it was just a steep, grassy hillside," says Mike Lucas. "There wasn't a plan for the house or landscape," other than a list of what the owners wanted—the number of bedrooms and bathrooms, a swimming pool, access to the stunning views, and placement of the house in the middle of the precipitous terrain.

"Looking up from the bottom of the site, I knew the project was going to take a lot of grading and engineering and a large number of retaining walls," says Lucas. "But once I got to the top, the view was breathtaking. I could see why the owners were set on having their house in that location." The view stretches across Sonoma Valley, overlooking the owners' guest cottage and the oak woodland below. On clear days, the San Francisco skyline is visible in the distance.

The driveway and arrival court were priorities. "Logistically, we had to get people up the hill—not only cars and people, but fire trucks, too, for fire building code," says Lucas. "Otherwise we'd never get a permit." To create an aesthetically pleasing arrival sequence, Lucas wound the driveway through the old oak trees that dot the property, maintaining adequate distance to protect their roots. Years later, the stately trees overhang the driveway in many places and cast gnarly shadows.

From the top of the hill, the driveway, which cuts across the property alongside large landscape walls, is nearly hidden from view. The arrival court, at the end of the windy driveway, is also out of view, set to the side of the house, preventing parked cars from disrupting the scenery. The view from the driveway captures glimpses of the stairs, stone and concrete walls, seating areas, and fountain. Ornamental grasses billow and sway in the breeze. Vibrantly blooming perennials add enticing highlights. "I wanted to greet people in a pleasing way and hint at what's to come," says Lucas. The prominent location of the stairs is functional. "Because the backyard and front yard were merged into one when the house was planned, we needed directional cues so people know where to go," says Lucas. "The stairs achieve this effect."

Lucas designed the stairs with mild-steel risers, concrete treads, and plantings that spill from planters on either side. "This is one of my favorite details of the project," Lucas says. The stairs reflect his design concept for the project—strong architectural lines blend with soft plantings, so the house and hardscape look like they are emerging from the hillside. "I wanted it to feel like the plants and other natural elements were slowly taking over," he says. Ground covers creep into crevices in the steps. Walls emerge from the hillside, and the lush plantings then transition into large tufa boulders placed to abut the tufa stone walls. Plants hang from wood planters.

Lucas was able to realize his vision for the project by working closely with the architect throughout the design. He also took the lead in developing the site plan. It included ten-foot-tall retaining walls behind the house as well as a cascading series of landscape retaining walls to shore up the hillside and make the home accessible by car and foot. "The walls were quite literally the building blocks of the project," says Lucas.

The materials palette of the home integrates seamlessly with the landscape. For example, both employ board-formed concrete walls, which Lucas used for retaining walls and to create a long reflection pool and tiered fountain. The architect's use of tufa stone, in a crisply cut rectangular pattern, is also echoed in Lucas's retaining walls. Farther from the home, where the landscape begins to transition to a more natural oak woodland, the dry-stacked walls terminate at large tufa boulders and are built in a random pattern that makes them appear more organic.

With so many walls, Lucas focused on ways to soften the hardscape. "Once the trails, stairs, and hardscapes were developed, the rest was a matter of infilling with beautiful romantic grasses and perennials that blow in the breeze and catch the light," says Lucas. Directly in front of towering walls, he planted large grasses, such as *Miscanthus sinensis*. Other deer-resistant grasses and perennials are planted lower on the hillside below the miscanthus. Groupings of grasses, such as *Pennisetum orientale* and Greenlee moor grass, sway in the breeze and catch the light, filling the garden with golden and tawny hues.

Swaths of perennials with brightly colored flowers, such as *Perovskia atriplicifolia* 'Blue Spire', *Achillea*

Architectural features along the approach to the front door—including a multilevel fountain that lines a set of steps planted with thyme and *Teucrium chamaedrys*—enliven the climb up the hillside.

The bocce terrace below the pool deck is bookended by two tufa-stone seat walls that were custom built by stonemason Edwin Hamilton. Hamilton custom cut, fitted, and finished the stone on-site.

'Walther Funcke', *Teucrium chamaedrys*, and *Penstemon* 'Midnight' add vibrancy to the otherwise earth-toned palette. Among the gaps in the hardscape around the steps are additional *Teucrium chamaedrys* as well as *Thymus* 'Victor Reiter'.

To minimize the need for handrails along the upper terrace next to the house, where much of the outdoor entertaining occurs, the design team placed the vanishing-edge pool at the edge of the terrace, incorporated wood planters along the retaining-wall perimeter, and added a reflection pool that's about three feet wide and more than twenty feet long. "These elements hold people back from the abrupt edge of the retaining wall in an elegant, unobtrusive way," says Lucas. They also offered a spot for artwork to adorn an open space next to the pool—three shell-like stone sculptures by stonemason Edwin Hamilton, who created all of the stonework on the property.

"These surprises pop up when you're open to them," says Lucas. "The walls, the reflection pool and fountain, the artwork—they were all an evolution of sorts. I didn't know the answers or what would develop when we began. One solution led to the next as the process unfolded and different pieces of the construction were completed."

This integration of artwork in the landscape can be found on the lower terraces too, where Hamilton custom built two finely fitted seat walls off either end of the bocce court. (The bocce court wasn't part of the original design, but as Lucas got to know the owners, he learned that they loved games. They developed this area together and created a destination.) The lower terrace acts as a bridge between the two ends of the home and has wonderful views of the valley and the house. Sitting on the stone seats among feathery plantings, "you feel immersed in the beauty of the land," says Lucas.

1. Entry Drive
2. Auto Court
3. Residence
4. Guest Wing
5. Office / Rooftop Garden
6. Pool
7. Fountain
8. Firepit
9. Bocce Court
10. Gravel Terrace
11. Planted Steps
12. Hillside Gardens

ADVICE FROM LUCAS

Plant the Steps

Plant steps with low-growing plants in order to tie the hardscape to the land in a soft, organic way. This is the first time I applied this technique, and it remains one of my favorite applications because of the way the plants, steel, and concrete come together to form the steps.

Initially, the builder was concerned that the steel, which forms the face of the step riser, would delaminate from the concrete and become a hazard. To eliminate this risk, steel rods were welded to the back of the steel riser and tied to the rebar, so when the concrete was poured, it fused with the riser. (The steel that extends beyond the concrete was also tied into the walls on either side of the steps.) The small planting beds on either side of the steps are filled with low-growing, low-maintenance plants, such as *Thymus* 'Victor Reiter' and *Teucrium chamaedrys*, that tumble over the edges of the steps and down the riser, creating the perfect balance between hardscape and softscape.

LEFT Japanese maples underplanted with Greenlee moor grass surround the driveway and parking area, where the sound of a nearby water feature is audible.

RIGHT The use of landscape walls was vital to developing flat spaces for outdoor living activities and walkways. Lucas used layers of wispy grasses and plants to ease the visual impact of the walls.

LEFT The slopes surrounding the bocce court form the perfect spot for textured ornamental grasses and colorful perennials, such as fluffy tan bristles of *Pennisetum orientale*, yellowish orange blooms of 'Walther Funcke' yarrow, dark pink flowering *Teucrium chamaedrys*, lavender colored 'Blue Spire' Russian sage, light purple 'Blue Fortune' giant hyssop, and dark purple 'Midnight' penstemon. Taller grasses—*Miscanthus sinensis* among them—obscure large retaining walls and accentuate the cantilevered concrete deck above.

RIGHT This retaining wall, located closer to the surrounding oak woodland than to the house, was faced with irregular lengths of linear tufa stone to create a more organic look. The wall helps ease the transition from the straight lines of board-formed concrete walls to the random stone pattern farther down the slope.

LEFT The view from the top of the hillside showcases wide-open vistas of the valley. The firepit was designed by Lucas and carved out of a large boulder. Sculptures on the pool deck are by Edwin Hamilton.

LOWER LEFT The long, shallow water feature provides a barrier between the patio and the top of a large concrete retaining wall, eliminating the need for handrails and enabling an uninterrupted view of the garden, oak woodland, and Sonoma Valley.

RIGHT In this spot on the top terrace adjacent to the home, a wood planter filled with *Chondropetalum tectorum* and *Senecio mandraliscae* at the edge of the pool deck forms a partial barrier. Where railing was necessary, its black color and minimalist design makes it practically blend into the backdrop while providing a nice place to take in the view.

CHAPTER FIVE

Old Meets New

A wide gravel walkway runs through the courtyard behind the house. London plane trees next to the home help create the proper proportions in the garden and also obscure the two-story-high walls of the house.

At an estate on a hillside overlooking the owner's vineyard in Healdsburg, California, Mike Lucas designed a garden that connects the old-world, stone-clad home with the greater landscape. When Lucas first saw the property, he found the scale of the two-story house and the materials selected for its exterior to be imposing and disconnected from the surrounding garden. He wanted to make the arrival sequence more welcoming and create a series of transitional spaces around the home that would help nestle it into the land.

"I've rarely encountered a driveway or arrival area that I liked from the start," says Lucas, "[and] this one was cramped and cage-like, running along an old wood-and-wire fence. Who wants to drive along that every day?" He moved the driveway more than forty feet to the middle of the property among grape vines that line both sides. The driveway winds past valley oaks that follow the contours of the hillside, and the house is revealed on the approach.

To soften the stone exterior of the home, Lucas planted one-hundred-year-old olive trees around a parking area. The olive trees fit thematically with the old-world structure and help transition from the large-scale walls of the home to the human scale of the garden.

The parking area, too, needed to be screened from view. In the arrival area, Lucas nestled the parking spaces below the front of the home, where the cars could be surrounded by plants and tucked out of view as much as possible. Tough, waterwise varieties, such as Lindheimer's muhly grass, rosemary, lavender, salvia, penstemon, germander, and moor grass, were planted beneath the olive trees and along the walkway that meanders toward the front door.

The winding walkway is key to an arrival experience that highlights the views of the vineyard and the land. From the moment the owners or guests step out of their cars, a series of stairs and landings form a scenic interlude. In one direction, there's a view over the vineyard and valley below. Another view focuses on a stone wall, majestic oak tree, and the owners' home brewery at the edge of the native forest. The entry path to the front door flows past a custom stone fountain that attracts hummingbirds and bees. The stone exterior of the house is softened with Boston ivy and plants that tumble over walls that line the driveway. "By the time you get to the front door, you've been welcomed by the land," says Lucas. "There are places to explore that provide topics for conversation."

A breezeway between the home and garage leads from the front yard to a large courtyard in the back, where outdoor entertaining occurs. The courtyard offered another opportunity to connect the home to the landscape. Lucas hired stonemason Edwin Hamilton to build a stone seat wall around an old, sculptural oak tree. "The stone has a gravity and architecture of its own," says Lucas. "It has visual weight and energy, so all the attention doesn't go straight to the home. It's balanced."

To soften the home's exterior and create shade in the courtyard, Lucas planted London plane trees in a grid. The trees continue along a gravel path that forms the central thoroughfare of the garden. Hornbeam trees trimmed into columns echo the pillars supporting the outdoor dining pavilion and lounging area. "Whenever possible, I try to extend architectural features of the home into the garden and bring elements of the land closer to the home," says Lucas. "Near the home, there's a merging of worlds. The design should reflect that." Tucked among the hornbeams is a modern reflecting pool that brings the sky, nearby oaks, and swaying grasses nearer to living areas. Birds frequent the shallow reflecting pool, bringing wildlife into living spaces.

Connecting the pool area to the surrounding landscape required careful consideration of the views to the vineyard and valley. "I didn't want the entire view to be seen all at once though," notes Lucas. "It sounds counterintuitive, but I feel it's better to frame the view with trees and break up one expansive view into segments." Working with the owners, he found places around the pool deck to plant olive trees and Chinese fringe trees to create a sense of depth. Unique views are revealed from various lounging, dining, and seating areas around the pool.

Large slab steps and boulders form a staircase leading from the pool down to the vineyard. Jerusalem sage, Santa Barbara daisy, and artichokes surround the steps and are tucked into gaps in the boulder retaining wall.

A set of steps leading down from a hiking path in the native forest behind the home leads to a bucolic seating area where the owners enjoy beer brewed in the structure on the left and wine made from grapes grown on the property.

Lucas also worked to soften the home's stone cladding and limestone decking in the pool area. "I wanted to break up the hard surfaces and add a softer, more relaxed feel in the space," he says. Boxwoods edge planting beds, and they are also planted in containers near the home in topiary forms to help extend the architecture into the pool area. Roses grow up the side of the home and sprawl from planting beds. Grasses, honey bush, dwarf olive shrubs, rosemary, Jerusalem sage, Mexican sage, Santa Barbara daisy, and germander are all loosely maintained to create a whimsical, less formal feeling around the pool.

"When designing, I'm often concerned with balance—old and new, traditional and modern, hard and soft, expansive and intricate, natural and architectural," say Lucas. "The landscape shines when the aesthetics and mood of each space are in balance." On this project, he worked with the owners to develop a garden that harmonizes with the surrounding landscape, vineyard, and valley—a place the owners (or their guests) would never want to leave.

1. Guest Parking
2. Residence
3. Garage
4. Brewery
5. Breezeway
6. Pool
7. Fountain
8. Sycamore Walkway
9. Sculptural Oak
10. Outdoor Kitchen / Pizza Oven
11. Vineyard

ADVICE FROM LUCAS

Aging a Landscape

Create a sense of maturity in a new landscape by using the right materials, techniques, and plants. On this project, walls were built using aged rugged tufa stone and hand-chiseled joints. A bowl-shaped fountain was carved from a solid block of limestone and fitted with a rustic wooden post for the spigot, and the limestone has a natural patina that ages beautifully.

By using strategic techniques while pouring board-formed concrete walls around the parking area, we were able to make them look old. We left gaps between the boards to create rough lines, and we used a drier mix and didn't seek perfection when filling the forms, which resulted in small holes on the surface. This technique gives the look of aged walls that have been there for decades: they looked like ruins upon pulling the form boards. Creeping rosemary cascades over the top edge of our aged walls to create a soft look.

Another way to age a landscape is by incorporating old trees. If there are existing trees, such as specimen oaks, that are already on the property, I emphasize their presence. Here, I brought in field-grown olive trees that are more than one hundred years old. The rugged trunks look ancient, and their size brings instant maturity to the landscape.

ADVICE FROM LUCAS

Thinking About Movement

Movement and transitions throughout the garden should feel seamless and effortless. For this project, one challenge was to clarify circulation. The existing route to the main entrance was confusing, so we directed foot traffic to the desired arrival location. We also added a continuous walkway along the front of the house to access secondary doors. In another instance where circulation was important, we emphasized stellar views from the breezeway that connects the front yard to the backyard by installing a long path through the garden that draws the eye to the other side of the property.

Circulation defined the entire site plan. It became the skeleton from which all else emerged, from the guest parking to the front door and beyond.

ADVICE FROM LUCAS

Softening Architecture

Tie the hardscape to the building to give the landscape structure, rhythm, and organization. Without something to tie the hardscape to the land, however, the hardscape can look stark or even out of place. Think about softening the architecture in terms of yin and yang. The walls and hardscapes are the bones of the garden (yang), and they need to be balanced by the soft, free-flowing looseness of the plants (yin) to harmonize the garden.

Plants can help settle buildings into the landscape. In this garden, we planted roses and Boston ivy at the base of the home's facade; they creep up the walls and wrap around downspouts—to soften the imposing stone exterior. I also used London plane trees in the backyard and olive trees in the front yard to create layers of greenery that give the landscape depth.

I like to plant ground covers, such as creeping rosemary, *Rubus pentalobus*, and thyme, that tumble over paving and retaining walls. Smaller perennials, such as Santa Barbara daisy and *Teucrium chamaedrys*, can be tucked into gaps in retaining walls or along steps and walkway edges.

ADVICE FROM LUCAS

Embrace Constraints

Constraints are a part of every project. Budgets, the availability of plants and other materials, construction schedules, and topography can all put limits on a project. Allow limitations to lead to creative solutions. For example, the considerably less expensive tufa stone steps at the entry to this home were installed because the eight-foot-wide stone slabs originally specified came in over budget.

Edwin Hamilton, the stonemason on this project, suggested that we obtain large tufa boulders that matched the walls we were building and have them cut to create the steps. The only caveat was that there would need to be a visible seam in the slabs to achieve the desired step width. The steps were completed within the original budget. The owners love the steps, seams and all—they are a true work of art.

ADVICE FROM LUCAS

Consider What's Underfoot

Think beyond the visuals when selecting the material for a path. The sound that a pathway makes as people walk over it influences how a landscape feels. A gravel path that crunches has the informal feel of an old-world garden or a native landscape, whereas a solid concrete path makes little sound when walked on, giving it a more formal, and purely functional, feel.

The materials and layout influence the pace at which people move through a space. A winding gravel path, like the front walkway designed for this project, creates a meandering approach, on which visitors can slow down and enjoy the scenery—vineyards and valley, trees and tall grasses, water and wildlife. It sets a mood that a straight solid path could not.

LEFT To break up the monotony of the stone pavers around the pool, Lucas created planters for boxwood hedges and roses. Larger conical boxwoods in terra-cotta pots next to the house add variety and a touch of Tuscany. Roses and ivy grow up the side of the home's stone walls.

RIGHT A lounge area by the pool overlooks the owner's vineyard. Roses, surrounded by boxwoods, add fragrance by the pool.

LOWER RIGHT Two hornbeam columns on grid with the breezeway help define the space surrounding the reflecting pool while extending the architecture into the landscape. The columns create a focused entry point to walk from the courtyard to a breezeway that provides shade for the pool area.

LEFT Views to the vineyards and rolling hills open up as one walks through the breezeway to the pool area. *Melianthus major* shrubs both frame and create the view.

LOWER LEFT A curved stone wall placed carefully around an old oak tree helps magnify the tree's presence in the landscape. Tucked along the edge of the forest is a brew house and beer garden.

RIGHT Most of the seating and dining areas are tucked under the covered breezeway, and a few smaller seating areas are placed around the pool to capture some of the most stunning views.

LEFT The arrival and parking area is tucked below a retaining wall and surrounded by one-hundred-year-old olive trees. Creeping rosemary cascades over the edges of the wall, while larger varieties of rosemary, 'Miss Jessop's Upright' and 'Collingwood Ingram', fill in behind the low-growing rosemary with perennials and ornamental grasses in back creating layers and depth in the planting bed.

RIGHT Boston ivy was planted at the base of this home's walls. ***Rubus pentalobus*** planted above a retaining wall next to the garage connects the wall to the landscape beyond.

This modern reflecting pool, a central feature of the courtyard, brings the sun, sky, leaves, and wildlife into the garden. A wood bench makes it easy to sit and dip a hand into the pool.

CHAPTER SIX

Treetop Hideaway

In the backyard, *Muhlenbergia dubia* catches the light from the evening sun. The pool extends beyond the hillside, which is supported by a gabion wall. Plants surround nearly every hardscape element, helping them merge with the land.

"There's magic when you work on a project with a general contractor who is also a master gardener," says Mike Lucas. He and builder Robin Guilfoyle of Robinwood Construction planted portions of the new landscape as the foundations were being poured for this residence in the hills of Healdsburg, California. "Planting early in the construction process meant that the landscape would be mature by the time the owners moved in," say Lucas. He also brought in forty-year-old olive trees and other large specimen trees and shrubs, preserved existing well-established trees, and planted grasses that mature quickly.

Following four years of design, planning, and construction, the owners, a couple from the Central Valley, moved into the family compound of their dreams—complete with lush, well-established trees and plants. "It was designed as a retirement home and family retreat," says Lucas. Several living quarters, designed to accommodate the owners' children and grandchildren, are spread across a one-and-a-half-acre section of the eight-acre site. The main challenge for Lucas's team was to connect the individual structures across a number of elevation changes, from a parking area at the top of a hillside to the bottom of the developed site, where the pool and cabana are located. Working closely with the architect, Lucas created a series of interconnected steps, landing areas, and paths that family members can enjoy as they traverse the landscape on their way to gatherings and other communal activities.

Seven structures dot the site, and numerous concrete and stone walls hold back the south-facing hillside. "The structures created opportunities for nooks, pockets, and long walkways between the buildings," says Lucas. "These intimate spaces balance the larger, wide-open views in other parts of the property." The landscape unfolds, which means that it's not experienced all at once. The owners and their guests make their way down the entry steps past olive trees to find a relaxing fountain near the front door. Or, if they turn a corner, an intimate seating area is revealed. "Surprises all over create a feeling of mystery, exploration, and excitement," says Lucas.

Near the upper parking area, olive trees and grasses frame the entryway to the main house and beckon people toward a series of steps and landings that lead to the various living quarters. The steps and landings are set into a slope covered in a tapestry of Mediterranean plantings. "This space is punctuated by old olive trees with sculptural trunks," says Lucas. "Swaths of plants, such as feather reed grass, artemisia, lavender, and Jerusalem sage, present beautifully when viewed from the front door or through windows of the house." The planted hillside also screens parked cars from view.

In a corridor that leads along the front of the house, Lucas placed a lush mix of plants in the shadows cast by the home. "This area reminds me of a mountain glen," he says. "The linear space looks and feels soft and welcoming as you approach the home." The plantings transition from a low-water palette of olive, ornamental grasses, verbena, and lavender to a verdant palette of ferns, Japanese aralia, and papyrus. "I enjoy thinking about how plants affect mood," says Lucas of his choice to often include water features, lush plantings, and seating in entryways. "These elements give visitors cues that they will be taken care of. There's water. It's relaxing. They can take a seat and simply enjoy being there. It affects people psychologically."

The walkways that lead around the sides of the home to a pool and the main outdoor living areas are bounded by a mix of black tupelos, ginkgo, lilacs, chaste trees, and perennials, such as Berkeley sedge, cardinal flower, Egyptian papyrus, and ferns. "The loose plantings balance the hard lines of the home and marry its architecture to the natural, living environment," says Lucas.

"There's also a can't-miss spot along the path," says Lucas, referring to a tree-house-like structure designed to capture views of the surrounding property. It's located where a small, dilapidated teahouse once stood. "It's really special and a favorite spot of the homeowners," adds Lucas. The outlook also serves as a counterbalance to a large tower that houses a garden shed, wine cellar, and playroom.

In the backyard, there's a small change in elevation from the home down to the pool area. "We experimented with different levels to get the feel right," says Lucas. The pool area needed to be lower to preserve the view from the house, but the team didn't want too many steps that would make it a chore to walk between

This wall, built with native stone, guides visitors from the arrival court to one of the living quarters on the property. Mature olive trees line the entry steps, greeting visitors with their outstretched trunks.

A large board-formed concrete retaining wall holds back the hillside between the arrival area and the house to create space for a walkway to the front door. Lucas installed a fountain in front of the wall to offer the soothing sound of running water and a feeling of respite.

the pool and the house. Lucas, along with the architect and homeowners, decided on an elevation change that required five steps. "It sounds like minutiae, but stairs can dramatically change the feel of a space," says Lucas. "I always put a lot of thought into how many steps to use, how wide they are, and how they feel when walking up and down—it should be an easy stroll and fit with the land."

At the base of the steps, tucked into one side of the pool area, is a sunken firepit, with a stunning view overlooking the pool, bocce court, and outdoor pavilion. To the west of the firepit, Lucas planted a large red oak, which serves multiple functions: a connection to the forest that surrounds the property, a transition in scale between the house and garden, and a source of much needed afternoon shade on the pool deck.

"Trees often have the most profound impact on the look and feel of a space," says Lucas. Throughout the property, he used trees in a number of ways: floss silk trees planted along a property line screen neighbors; ginkgo trees near the overlook pay homage to the old teahouse; and olive trees screen cars and add a sculptural quality. Tupelo trees in the entry glen soften the architecture and draw visitors down the lush path; their green, oval-shaped leaves turn scarlet red in the fall, providing seasonal appeal. "Trees are like beacons," says Lucas. "They provide context and call people farther into the landscape to explore."

Adjacent to a large oak tree that was preserved, Lucas and the architect tucked in a large, open-air pavilion, which is outfitted with an outdoor kitchen and barbecue, seating, a TV, heaters, a fan, and a refrigerator. "You could practically live in there," says Lucas. Roll-up screens between the pillars can be adjusted to provide shade throughout the day. The pavilion is the central hub of the backyard, offering views of the surrounding forest as well as proximity to various spots where the family can lounge or play bocce without feeling separated from the party.

These connections make the landscape shine. "There's an artistry to it," says Lucas, who often looks for ways to build connections between spaces and among structures. This is often facilitated by developing a strong working relationship with the architect or designer. "There's a fine balance in ensuring that each element gets their due, whether it's living plants or architecture, walls or wildlands," says Lucas.

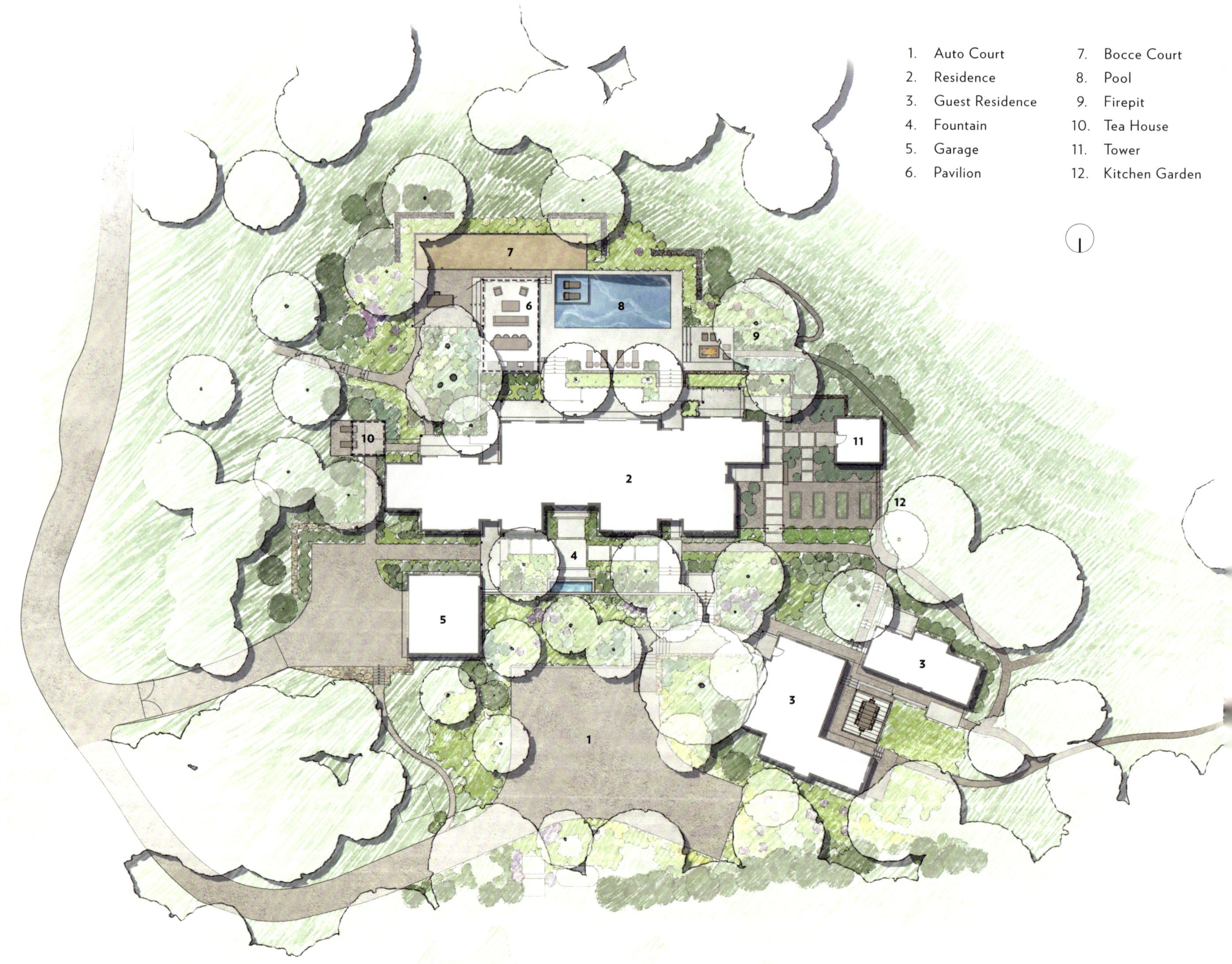
1. Auto Court
2. Residence
3. Guest Residence
4. Fountain
5. Garage
6. Pavilion
7. Bocce Court
8. Pool
9. Firepit
10. Tea House
11. Tower
12. Kitchen Garden

ADVICE FROM LUCAS

Know the Land

Be aware of how a site feels. Get to know how it impacts your energy level and mood. In order to create a garden design that responds to the land, I spend time walking the site, sitting under trees, and exploring different areas at different times of day to gain an awareness of how the sun, shade, breeze, wildlife, views, or other elements make me feel. I listen to the birds. I feel the breeze. I watch the sunset. I imagine the grasses swaying in the breeze and catching the light.

As I experience this, I gain a better understanding of where to place a seating area, pool, path, or other architectural elements that will allow the owners to experience these same feelings. As a landscape architect, it's my responsibility to help my clients discover the special places on their land and make them more accessible, enjoyable, and comfortable.

On this project, the owners were really drawn to an old teahouse perched on the edge of a slope that offered wonderful 360-degree views of the surrounding forest. Unfortunately, the structure was unusable, and the design team replaced it with a modern "tree house." The owners love this spot and its close proximity to the main bedroom.

LEFT The doors on the back of the house open to views of the surrounding forest and sky. Lucas kept this view uninterrupted by installing a thirty-inch retaining wall that lowered the elevation of the pool area to limit the visibility of the pool, fire feature, and pool furnishings.

UPPER RIGHT A "tree house" was placed on the edge of the slope where an old teahouse once stood.

LOWER RIGHT The view down the entry courtyard is lush with plantings. Strong architectural lines are blurred as grasses and shrubs overflow from their planters. Berkeley sedge tumbles over the edges of the pathway, and Corsican mint creeps along the gaps in the walkway, further softening the hard lines of the concrete paving.

LEFT The steps from the pool to the guest rooms are connected visually with plantings that surround each structure to help knit the spaces together. Yellow blooming Jerusalem sage, rounded Himalayan sweet box, and Berkeley sedge fill planting beds. A black tupelo tree provides shade in the hot summer months and turns a vibrant orange color in the fall.

RIGHT A firepit with built-in seating was constructed with poured concrete, giving it a sculptural quality. Lucas sited the pool and firepit lower than the house to ensure an uninterrupted view from the house. Much-needed shade is provided by the wood-paneled outdoor pavilion and the red oak next to the pool.

LEFT On this landing at the bottom of the slope from the arrival area, the entry courtyard is fully visible while the cars are hidden from view. Custom metal railings painted a dark gray coordinate with the finishes on the house and other structures. Midway up the steps, the path splits off to the guesthouse.

BELOW LEFT Olive trees were planted as construction on the home was starting to bring an instant sense of maturity to the landscape. They were paired with quickly maturing perennials, such as autumn moor grass and Spanish lavender, that would fill the spaces below the canopy.

RIGHT The long trough fountain and three large scuppers accent the concrete wall. At around twenty inches high, the basin's edge is designed for sitting, where the owners can take in the soft, lush plantings along the entry walk—papyrus, Berkeley sedge, Renga lily, cardinal flower, Jerusalem sage, and bog sage.

LEFT A bocce court runs parallel to the back of the pool and next to the main lounge area and outdoor dining space, so players feel like they are part of the action.

RIGHT A lounge area in the outdoor pavilion offers views of the property in nearly every direction. The pavilion is fitted with adjustable shades, heaters for cool evenings, and an outdoor TV that descends from the ceiling.

CHAPTER SEVEN

Prairie Leisure

Lucas defines outdoor rooms using subtle changes in elevation. Just two steps up from the pool area is a dining terrace, set among old olive trees, that extends into the landscape like a peninsula and terminates at a fireplace.

For this ten-acre residence near downtown Sonoma, California, Mike Lucas developed a garden that would serve as home base for a couple from Texas (although originally from Canada), who wanted to retire in an area known for its food and wine culture. "The owners fell in love with this big flat property because of the open sky and prairie-like feel that reminded them of where they grew up in Canada," says Lucas.

These were the jumping-off points for the project—wine and food culture, great weather, family gatherings, and meadows that would elicit memories of the prairies the owners love. "Having worked in this region before, I knew the breezes and evening sun would work perfectly with ornamental grasses," says Lucas. "They add an element of movement, light, and dynamics that were missing in an otherwise plain, flat site."

The owners were also interested in developing a retreat where they could entertain their kids when they visited from San Francisco. The property included spaces for outdoor dining, a fireplace, an updated pool, an open lawn for games, as well as a bocce court and pizza oven. Lucas used these elements to create focal points in the landscape and to add dimensionality to the property. "Adding elevation changes, seat walls, structures, sculptural trees, and landscape walls was essential to developing the character of the property," says Lucas. "When we started it was just a big grassy field." Over a three- to four-month period, Lucas worked with the owners to turn the field into a prairie-inspired wonderland.

"Originally, the entry to the property was unassuming," remembers Lucas. "Being that we're in Sonoma, the owners had a vision of driving through a vineyard." Lucas connected them with a vineyard manager who grows grapes on estate properties. (The homeowners receive a small portion of the wine produced, and the vineyard manager harvests the grapes in return for his work planting, managing, and harvesting.)

"The vineyard transformed the arrival," says Lucas. "It winds through rows and rows of grape vines that are lush and green much of the year, covered in grapes through the summer, and vibrant orange and red in the fall." The driveway ends in a small gravel parking area in front of the home, where he added three landscape walls constructed with the same locally quarried stone used for the house's cladding. The walls ease the transition from the flat land that the vineyard is planted on to the two-story house with a tall, steep roofline. "It's less abrupt," says Lucas, who added a number of transplanted trees from the original garden at the front of the house for this same reason. "There's a lot of house."

The walls also screen out the parking area. "Since the owners disliked formal hedges, I used five-foot-tall walls to hide the cars," says Lucas. "The walls also balance well with the soft wispiness of the meadow. The grasses play perfectly off the rough stone."

Next to the parking area, a trough fountain welcomes people as they arrive and beckons them toward a gap in the walls that leads to the front door. The owners, wanting a casual and informal feel, decided on a gravel path leading to the front door. The path passes through a billowy meadow planted with fine-textured grasses, such as prairie dropseed, pine muhly, and Lindheimer's muhly. Other perennials such as Russian sage, shrubs such as dwarf olive, and strappy-leaved grasses such as 'Canyon Prince' wild rye help vary the textures in a sea of feathery grass blooms.

Around the side of the house, Lucas focused on dressing up the motor court and two garages. Next to the freestanding garage, which the owners turned into a ceramics studio, Lucas designed an extensive vegetable garden and orchard with a dozen fruit trees. The raised beds were constructed using leftover material from the paving at the back of the house—a strategy that has become a go-to for Lucas.

"The orchard and raised-bed vegetable garden have a natural feel and were designed to extend into the vineyard," says Lucas. "It gets the owners out among the vines, where they are immersed in the vineyard instead of viewing it from a distance. They can look down the rows and see the grapes growing on the vines."

Off the back of the house, the central feature is a large stone peninsula that extends out into the property at the same grade as the house—roughly one foot above the pool and several additional feet above the rest of the property. "The peninsula floats above the rest of the property to pick up a better view," says Lucas. "You're looking down on the meadow where grasses gently sway in the breeze or sunflowers bloom in bright yellow hues." At the end of the peninsula, a fireplace and lounge area cap the space. There was a little pushback about placing the fireplace where it would block some of the view to the west, but Lucas knew it would enhance the vista by creating a sense of depth and draw people farther into the space.

Two landscape walls screen parked cars from view and separate the private landscape from the driveway and vineyard that surrounds much of the property. The same stone was chosen for the walls as was used on the house cladding, which ties the architectural elements together.

Groups of perennials and ornamental grasses create a bucolic meadow in the backyard: *Achillea filipendulina*, *Cistus* x *hybridus*, *Miscanthus transmorrisonensis*, *Muhlenbergia dubia*, *Perovskia atriplicifolia* 'Blue Spire', *Pennisetum spathiolatum*, *Phlomis russeliana*, *Sporobolus heterolepis*, *Salvia azurea*, *Helianthus maximiliani*, *Verbascum bombyciferum* 'Arctic Summer', and *Sesleria autumnalis*.

On the peninsula, a large dining table is tucked away among six large olive trees with gnarly trunks and a canopy of branches that provide shade. "The fireplace and olive trees create views that change as you move to different areas of the space," says Lucas. "Instead of one big view you see from everywhere in the backyard, there are framed vignettes that can be appreciated individually." In one direction, Adirondack chairs sit under trees next to the bocce court. Another view reveals the pool and unique grading. In another direction, the swath of meadow is most prevalent.

Beyond the pool, where the owners were adamant about keeping a lawn, Lucas graded the soil to create a land sculpture type of aesthetic. Just beyond the existing swimming pool, a slope drops off a few feet to the level of the vineyard and meadow. "It's a simple thing to do but changes the visual dynamics of the property," says Lucas. "Looking out over the grasses that light up in the sun to the land and sky beyond is one of the things the owners love most about the landscape—the elevation changes are what make this possible."

Where the lawn stops, a vast crescent-shaped meadow begins. "There must be around a thousand plants in the meadow," says Lucas. "I chose plants that would dance in the Sonoma breeze and really glow." Perennials such as prairie dropseed, slender veldt grass, pine muhly, autumn moor grass, rockrose, Jerusalem sage, and Russian sage mix together in large mass plantings. "A real star out there is the Maximilian sunflower," notes Lucas. "It probably grows to eight or ten feet tall and explodes with bright yellow flowers late in the summer."

On the other side of the peninsula, Lucas sited the bocce court, which was on the owners' wish list from the beginning of the project. It's tucked in between the steps of the peninsula and a small seat wall. "The placement of a bocce court can be tricky," cautions Lucas. "It has to be in the entertaining zone of the backyard, or it never gets used." Here, the court is next to lounging, cooking, and dining areas.

Much like the kitchen triangle that's used when designing functional layouts for indoor kitchens, the bocce court, dining area, and pizza oven create a similar type of proximity for entertaining. "Everyone can still have a conversation because architectural elements, and thus the entertaining that goes on near them, are in close proximity—people can easily saunter from space to space or talk to someone playing while sitting at the dining table. Wherever people go, they are part of the action and nestled safely among the comforts the landscape brings—shade, seating, warmth, food, and scenery." ☼

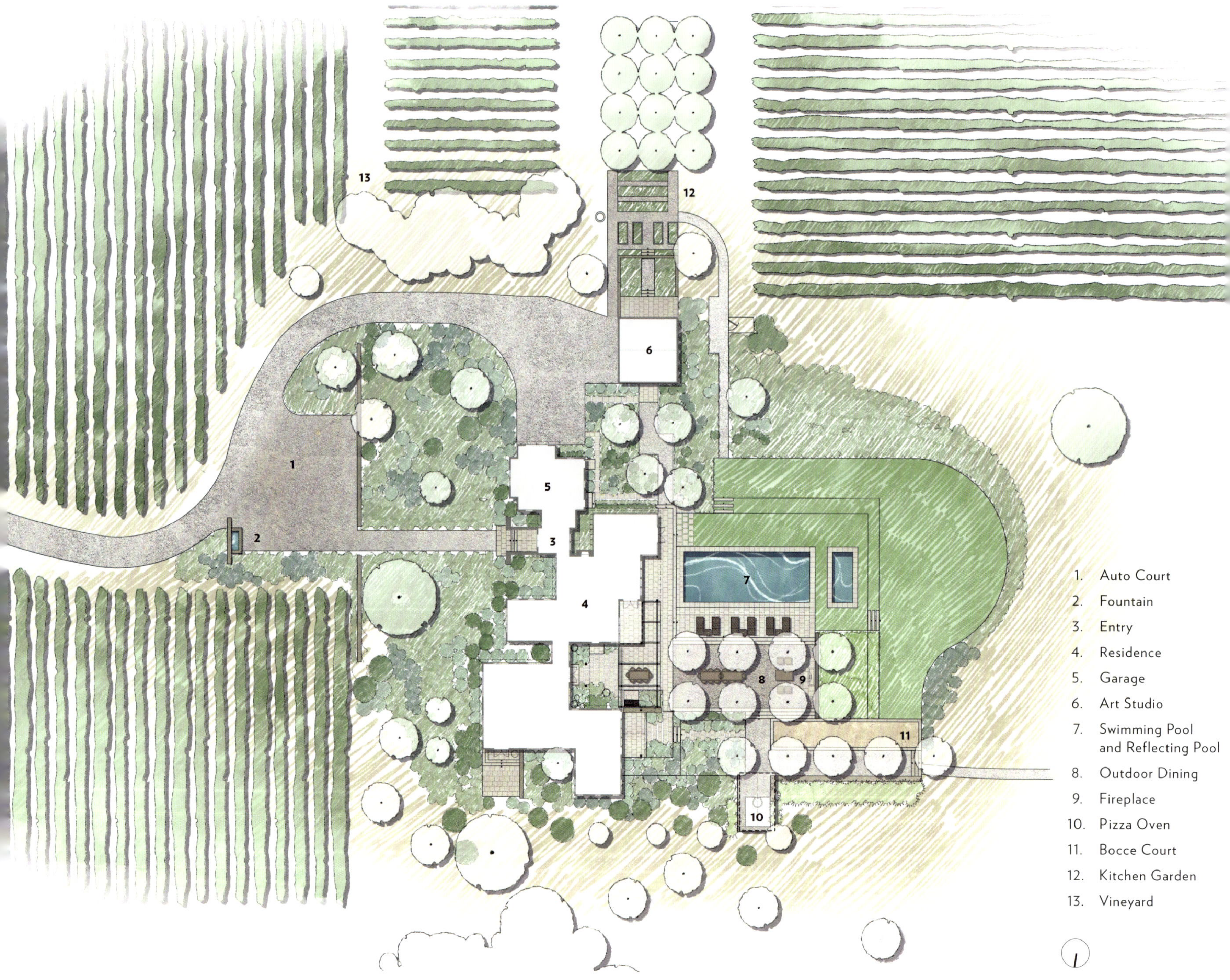
1
2
3
4
5
6
7
8
9
10
11
12
13
1. Auto Court
2. Fountain
3. Entry
4. Residence
5. Garage
6. Art Studio
7. Swimming Pool and Reflecting Pool
8. Outdoor Dining
9. Fireplace
10. Pizza Oven
11. Bocce Court
12. Kitchen Garden
13. Vineyard

ADVICE FROM LUCAS

Don't Fear Changing Course

Designs often continue to evolve during the construction process as the hardscapes are installed. The owners are also able to better understand their landscape as paths and walls begin to take shape and materials are brought in. They may have new ideas once they see the structures going into the ground.

Changes during construction often have the potential of becoming a cornerstone element of a design. On this project, the pizza oven and structure weren't part of the original plan. At the end of the construction process, the owners showed me a picture of a trip they took to Italy, with people playing bocce, drinking wine, standing next to a pizza oven. They said, "I've always wanted to have this."

As I considered the idea of adding the pizza oven, I noticed that our bocce court felt a bit lonely. Looking at the garden with this concept in mind, I realized there was a large empty space beyond the bocce court that could easily hold an oven. Instead of adding a basic pizza oven to the grilling area, I decided to build a rustic timber structure that would draw people to this side of the garden. Architecturally, it provides a magnetic anchor for the garden that was absent in the original design.

LEFT In addition to grasses, the meadow is also dotted with a variety of shrubs, such as *Rosa rugosa* 'Alba', *Olea europaea* 'Little Ollie', and *Frangula californica* 'Eve Case'. These shrubs bring form and structure to the layered mix of grasses and perennials, such as blue-foliaged 'Canyon Prince' wild rye, purple-flowering *Perovskia atriplicifolia* 'Blue Spire', and prairie dropseed. Several ash trees, which separate the front yard from the backyard, were transplanted from another area of the property.

LOWER LEFT This informal seating area to the side of the bocce court is tucked next to an outdoor pizza oven. There's shade, it's close to food, and there are great views of the property—amenities that make this space a destination within the garden.

RIGHT This fountain, made of stone that matches the house cladding, greets visitors when they step out of their cars. The ceramic inlaid panels that house the spouts are works of art by the owner and represent the bounty of the land.

LEFT The gravel walkway to the front door is set within a meadow of mixed plants, which includes pine muhly, Lindheimer's muhly, *Perovskia atriplicifolia* 'Blue Spire', prairie dropseed, as well as other flowering perennials that bloom in various seasons, popping up between the grasses.

RIGHT Lounge chairs on the pool deck overlook a tightly planted meadow border and the owner's vineyard beyond. A Bolero sod lawn with gopher wire is planted in front of the meadow. Layers of planting draw the eye to the horizon and sky.

FAR RIGHT The red-tinted inflorescence of prairie dropseed catches the sunlight perfectly in the morning and evening. These grasses were planted in large swaths throughout the front yard meadow.

LEFT Lucas used unique grading around the pool to add sculptural form to the expansive lawn on the otherwise flat site. The lawn rolls down several feet to the level of the meadow located along the edge of the lawn. The pool area, in front of old field-grown olive trees, . is elevated to provide attractive views of the meadow and vineyard.

RIGHT The pathway leads from the driveway to the backyard near the remodeled pool. Five multi-trunk 'Natchez' crape myrtle trees were planted around the path and underplanted mainly with blocks of French lavender hybrid, 'Walker's Low' catmint, and *Salvia nemorosa*. From there, the entire meadow can be seen, along with the rest of the backyard. The building in the distance is a garage that the owners converted into an art studio.

LEFT Leftover stone from the paving was used to build the raised beds in this lush kitchen garden, which is near the owner's art studio. A wood-and-wire fence keeps rabbits and other critters out of the garden, which is filled with vegetables and flowers and is connected to a small orchard of fruit trees that extends into the vineyard of productive grapevines.

RIGHT A path from the master bedroom to the bocce court and outdoor living spaces passes through an intimate garden. *Lavandula angustifolia*, *Lavandula* x *ginginsii*, *Salvia nemorosa*, 'Sally Holmes' roses, and *Olea* 'Little Ollie' surround the small walkway. Chinese fringe trees that line the right side of the bocce court provide much-needed summer shade and reinforce the structure.

This gravel dining terrace, which is at the same elevation as the house, extends into the garden. A large fireplace and lounge area hide the view in one direction. To the left is the pool and the view over the meadow; to the right is the bocce court and pizza oven.

CHAPTER EIGHT

Traditional Estate

A path around the house opens up to an outdoor kitchen and dining area fitted with a wide fireplace. Upright sycamore trees were chosen for their fast growth rate; an existing apple tree overlooks pink roses that add a pop of color.

When the homeowners first moved to this property in Healdsburg, California, the land was a large, flat lot with an existing traditional two-story main residence surrounded by dilapidated wood retaining walls, an aging pool, a number of rotting apple trees, and a massive dying oak. But it wasn't a lost cause. There were some wonderful elements to preserve and enhance: a vineyard that wrapped around the home, some beautiful crape myrtles and magnolias, and a few salvageable apple trees. The landscape work began with clearing out much of the site, leaving Mike Lucas with a blank slate. The goal was a new productive landscape focused on outdoor entertaining, relaxation, and growing food.

Lucas's main challenge was to create a garden plan based around the owners' ambitious wish list: a garage, an art studio, a kitchen garden, a pool house, a bocce court, an outdoor kitchen and dining space with fireplace, and an updated pool, as well as an abundance of plants. In addition, the property was very flat and there were no grand views. "We had to create all the views internally by developing paths, installing mature trees, and building architectural features, such as fireplaces and fountains that captivate people's attention," says Lucas.

Lucas started the design where he often does—with the driveway. "It's important to sort out the arrival," says Lucas. "It sets the tone." Upon entering the property, guests emerge from a stand of redwoods into a parking area that faces the home. Sculptural trunks of crape myrtles act as greeters. Masses of autumn moor grass fill the space between paths, and rows of boxwood and Mexican orange wrap around the home. The arrival is a lush garden setting, and there's a clear path either to the front door or to a walkway that splits to wrap around both sides of the home and into the backyard. This driveway was designed with space for visitors to get out of their cars, look around, see the vineyard, and enjoy beautiful plants in various shades of green.

Around back, smaller paths branch out from the main walkway into various garden rooms. The southern path—which leads to an art studio, a kitchen garden, and a small orchard—is defined by a variety of flowering plants. Roses grow over arbors and along the house, and mixed perennial borders feature nepeta, agastache, lamb's ear, Grecian horehound, verbascum, sedum, kniphofia, erigeron, and yarrow. The paving reflects the informal mood: stepping-stones are surrounded by low-growing thyme and oregano. It's a path meant for moseying and taking in the flowers and their fragrance.

The northern path, which leads to an outdoor dining area and the pool house, features a small carved fountain, a view of the vineyard, and masses of hydrangeas. Suitcases easily roll on this path's continuous paving—convenient when family and friends stay in the pool house, which doubles as a guesthouse.

The owners wanted to preserve the existing views of the vineyard from the back of the house. "Originally, I had planned to put the outdoor dining area and fireplace off the back of the house so it would be close to the indoor kitchen," remembers Lucas. "But the owners asked about moving them so they could see down to the pool and vineyard." Lucas liked the modification, and he moved the outdoor kitchen, dining area, and fireplace to the side of the property and sunk the outdoor kitchen and bar into the ground several feet down to further remove it from view. Lucas also placed the pool house and other architectural elements on the sides of the property to keep an unobstructed view of the lush gardens and pool.

"The site is pretty flat, with the house built up a few feet above the existing grade, so I wanted to draw out that change in elevation as much as possible," says Lucas. "Small walls, steps, and landing areas delineate separate spaces, where the garden experience can unfold in layers." Rather than having one large space around the pool, Lucas created an elevated terrace for lounge chairs. The landings break up the space, leaving room for lush plantings, plinths, and trees, and their slight elevation affords a better view of the pool and vineyard.

The pool was a holdover from the previous owners, but the new owners completely remodeled it. "We flipped it around," says Lucas. Originally, the steps had been on the opposite side, but the owners wanted to enter the pool next to the pool house. "I think we used a few walls of the original pool, but it was a complete redig and expansion," remembers Lucas.

He also had to develop a system to hide the pool equipment. The homeowners and Lucas love the solution:

The wraparound porch steps down to the landscape in several directions for easy access to outdoor entertaining spaces. Sorbet-colored roses add vibrancy to an otherwise monochromatic palette.

An existing pool was completely reconfigured and remodeled. Steps were moved to the other side of the pool for easy access from the new pool house, and terraces were constructed using small retaining walls, creating an elevated spot for loungers that enhances the view over the pool, bocce court, and surrounding vineyard.

a fountain, with water that spills into a large trough. The walls of the fountain conceal the equipment and soften its noise. "Sometimes these large projects are a little bit of trial and error until you find the best approach," says Lucas. "You have these little challenges that pop up everywhere. If you approach them with an open, collaborative mind, they lead to fantastic solutions."

The various garden rooms coordinate with each other, but Lucas says he wanted each to have a distinct identity, with "a hierarchy of scale with the trees." Each space has a main tree variety that matches the specific character of that space. The pool is surrounded by magnolia trees that will provide filtered shade as they mature. In addition, the magnolias' large leaves are easy to gather, so the pool filter is less likely to get clogged. Around the outdoor kitchen and dining area, fast-growing sycamore trees provide shade and verticality. Along transitional walkways, ornamental cherries bring overhead flowers to the garden in spring. The bocce court is characterized by old olive trees that the owners love because of their Italian heritage. In the kitchen garden are espalier fruit trees and traditionally grown figs, persimmons, citrus, and other assorted fruit trees.

The kitchen garden was developed toward the end of the project, after much of the garden was already built. "I assumed we were going to build some wood boxes for growing vegetables," says Lucas. As it turns out, he was able to build raised beds using extra stone from the walls in the pool area, the fountain, and the outdoor kitchen. These eight raised beds are intensely cultivated, while a few additional large beds at ground level are filled with pollinator-friendly plants and artichokes. Located at a frequently used intersection between the owner's home-office space and the detached art studio, the kitchen garden is one of the owner's favorite spots.

Lucas seeks inspiration everywhere. "I like to look at each space within a garden and try to infuse it with its own personality and purpose," he says. Lucas works throughout the design and construction process to improve views, to make spaces more accessible or more comfortable, or to find special architectural elements, such as fountains, to enhance a path or patio. "I love addressing these challenges, finding cool solutions, and then putting everything together into a cohesive design that fills the owners' lives with joy," says Lucas.

1. Auto Court
2. Garage / Art Studio
3. Residence
4. Fountain
5. Pool House
6. Pool
7. Spa
8. Bocce Court
9. Outdoor Fireplace
10. Outdoor Kitchen and Dining
11. Water Feature (Pool Equipment)
12. Vineyard
13. Kitchen Garden

ADVICE FROM LUCAS

Make Transitions Deliberate

Punctuate transitions with plants, containers, or boulders to celebrate the movement from one space to another. Pay particular attention to grade changes, where steps can enhance the sense of relocation between outdoor rooms.

This project had mostly flat topography, although the home was originally built two feet above an existing swimming pool. We needed to remove a failing wood retaining wall; subsequently, we installed small terraces and steps. Each step has broad twenty-inch treads and low four-inch rises to create a gradual, easy descent from the house to the pool. Three paths lead down from the house level, and each has the same grade changes. The wide stairs and an extended terrace convey relaxation—there's no rush for the experience to unfold.

LEFT This path links the front yard and backyard as well as the vegetable garden and house. There were no natural views from the garden, so the scenery along this path was entirely constructed through the use of landscape structures, plants, and paving.

RIGHT A spa in the corner of the property perches several steps above grade, overlooking both the vineyard and the flower-filled garden.

LOWER RIGHT The walkway around the home looks different in each season as various plants come into bloom. In summer, the hydrangeas hit peak bloom along this path. Sunlight streams in through a canopy of crape myrtle branches that provide enough shade for these big-bloom beauties.

LEFT Lucas used slight elevation changes of just a few steps to create outdoor rooms in the garden's original large, flat, open space. Boxwood hedges surround these formal-looking lounge areas between the pool and bocce court, while 'Little Gem' magnolias, with a pyramidal form and large, easy-to-clear leaves, provide shade for the pool.

RIGHT One of the owners is Italian and requested a bocce court. It was placed at the back of the property, along with a small patio with wonderful views through the garden, toward the house.

LEFT Irises, asters, geraniums, roses, and hydrangeas border this paved path around the north side of the home. A worker at the stone yard carved the welcoming fountain's stone bowl.

LOWER LEFT Lucas planted 'Potomac' crape myrtles around the arrival and parking area to connect the new landscape with large existing crape myrtles. He chose a monochromatic color palette for the plants that surround the porch, using shades of green with varying textures to avoid overpowering the white house.

RIGHT During construction, the team added a wall fountain and trough that conceals the pool equipment.

The kitchen garden beds were built with leftover stone from the paving and walls. An informal wood-and-wire trellis supports espalier fruit trees and separates the kitchen garden from the rest of the garden. The owners' art studio, which is connected to the garage, is visible beyond the raised beds.

CHAPTER NINE

Native Refresh

A sparse palette juxtaposed with the wild backdrop allows trees to appear prominent. A structure that houses a bath, a sauna, a changing room, and a storage area nestles into the hillside, out of the main views. Its modern style coordinates with the house, and its dark gray color blends into the background.

Deep in the densely wooded hills between Santa Rosa, California, and Napa sits a midcentury-modern weekend retreat. The owners—a family of four—travel about sixty miles from San Francisco to get there, driving up and down hilly dirt roads, winding and meandering through Douglas fir forests and oak woodlands. "It's really remote," says Mike Lucas.

When the family came to Lucas seeking to install a pool, he knew that in order to turn the property into the weekend retreat the family wanted, parts of its wildness would need to be tamed. "It needed the creature comforts to balance the rugged beauty that was inherent in the site," says Lucas.

When he started working on the project, the home was still in the design stage of a remodel. The couple was working with an architect to convert its old post-and-beam construction into an open floor plan that would take full advantage of the stunning views of the forest and distant hills. "It's perched on the side of a ridge," says Lucas. "Driving up the dirt road, I knew whoever had purchased this place had to have an adventurous side."

"After an early visit to the property, I felt one of the best spots for the pool would be the side of the hill overlooking the valley," says Lucas. But he knew that this location would require substantial engineering, with retaining walls to create a usable flat area around the pool. He had observed that although large glass windows and doors at the back of the house provided wonderful views, there was no reason to go outside into the backyard. The couple, the architect, and Lucas all felt that siting the pool close to the home would entice the family and their friends to spend time outdoors. It would also take full advantage of the property's unique views. "The alternative, less ambitious, plan for the pool had it located in an open valley at the bottom of the knoll," Lucas remembers. "But that put it too far from the house. It didn't enhance the landscape, the view, or the house."

The owners decided on the upper location. "Though this option was more expensive, it's the one that created drama and excitement," says Lucas. He located a sauna, changing room, barbecue, and shower off to the sides and lowered them partially below grade so these more functional elements wouldn't interrupt the dramatic views. "It was vital to preserve the open vistas and keep the clutter out of view," says Lucas. The pool and landscape walls are also sunken below the house to further expand the open views and maximize space. This location is easily accessible from the house, and the kids can swim in the pool while in view to those inside.

The pool serves as the focal point of the backyard, beautifully reflecting the sky, rolling hills, and nearby trees. Long, low walls built into the grade help nestle the structural landscape features into the hillside. The materials used for the concrete walls and pool deck are modern, but the colors are understated earth tones.

"This property was all about balance," says Lucas. "The modernity of the architecture balances nicely with the rugged native land and plants. Near the pool, two giant coast live oaks were preserved to frame views of the forest and also to create gorgeous shadows in the morning and evening; sunlight beautifully backlights them at various times of day. "Tucking the pool among the oaks had surprise benefits," says Lucas. "I love seeing connections between plants and architecture, such as where the oak branches grow over the corner of the pool."

Once the pool and outdoor entertaining spaces were established, Lucas focused his attention on drawing people into the landscape. He began with the arrival area around the driveway. An old garage was removed to create an open parking area under the native oaks, away from the house and views. Gravel paths and stone retaining walls greet the family and visitors and lead to the front door of the home. One path leads to the front entry court while another leads to the backyard, where a more intimate seating area and the pool await.

The path to the entry court and front door is informal. It's studded with lichen-covered boulders, naturalistic plantings, sculptural oaks, and gravel paving. "There's a simple beauty to the walkway—you hear the crunch of the gravel underfoot, you see a small architectural seat wall up by the door, lizards dart about, you feel the breeze washing through the oaks as grasses sway back and forth," says Lucas. "It creates an inviting arrival experience that blends with the native woodland." The path to the backyard has a similar feel—gravel underfoot and soft natural plantings. These plants are all native to the surrounding hills: manzanita, native heuchera, coffeeberry.

The path to the backyard emerges under one of the large live oaks overlooking the pool and a small seating

Boulders mark the entry to a narrow path that invites exploration into the more rustic areas of the property.

This view from within the front entryway has a Zen-like quality. A concrete retaining wall holds back the landscape from a steep slope, creating a flat, open space for arrival and departure. This is balanced with lichen-covered boulders installed within the courtyard to soften the architecture and the craggy tree trunks beyond.

area on the side of the hill. "I love looking for incredible views and then finding ways to make them more accessible," says Lucas. "This is one of those spots—right off the back of the house—where the beauty of the land and architectural landscape all come into view. It really hits you."

Lucas found that once the hardscapes were built, the landscape felt angular. The hard lines needed to be blended, which he achieved by using plants. He installed ground covers, such as creeping raspberry, under the oaks because they require little water. Grasses and grasslike plants blend well because they are similar to plants that were already growing in the native landscape. Lucas used native perennials, such as *Symphyotrichum chilensis*, 'Salmon Beauty' yarrow, and Douglas iris, for pops of color among the grasses. Nearer to the house, Lucas interspersed a few flowering shrubs, such as Western redbud, to soften corners. "The plants help tie the new landscape to the native land," says Lucas. "They also blur the edges of what could be stark architectural features, such as walls, pool decks, or steps."

Under the oaks and around the front of the home, Lucas followed a wild planting style. Around the pool, where the architectural lines are more formal, he planted large masses of ornamental grasses, opting for grass and grasslike plants such as Atlas fescue and Santa Barbara sedge on a grid near the home and in planters above the concrete retaining walls. In addition to connecting the built environment to the property beyond, the plantings are deer resistant, low water, and low maintenance.

"The goal was to civilize the wild property—make it comfortable and add some flat spaces for outdoor living but otherwise stay out of the way," says Lucas. "I focused on enhancing the elements of the site that were already there—the sky, breezes, tree shadows, views, the sound of birds." Any elements that were necessary but not additive were hidden from view. "I wanted to preserve the treetop getaway feeling while making it comfortable for family retreats."

1. Auto Court
2. Entry Walkway
3. Covered Entryway
4. Residence
5. Pool
6. Pool House
7. Pool Patio
8. Barbecue Area
9. Overlook

ADVICE FROM LUCAS

Designing for Shadows

Design hardscapes to showcase tree shadows. I set the pool in the property's sunniest spot, between two large oaks that frame great views of the native forest and rolling hills. These trees are so close to the pool and patio that the branches overhang in spots. When the morning or evening sunlight streams through the trees, it creates gorgeous, intricate shadows on the surface of the smooth concrete pool deck—a pleasant surprise that I hadn't planned for.

If I had gone with a rough stone, placed the pool deck farther from the trees, or had plants in these areas instead of smooth concrete, the effect wouldn't have been the same. In times like these, when I get lucky, I try to analyze the results to apply in future projects.

ADVICE FROM LUCAS

On Framing Views

Consider big views thoroughly and treat them with respect, but do not make them overly precious. Often, covering up part of a view helps the design. If there's too much going on, it's difficult to know what to look at.

A tree can frame a magnificent view. Though it may take some of the vista away, it can also make it more dynamic, as objects in the foreground help create a sense of depth. Architectural elements, such as walls, can similarly shield views in some areas and reveal them in others.

I sited this pool between two incredible coast live oaks that frame the view to the hills beyond. This placement creates a relationship between the view, the trees, and the built landscape: the trees, sky, and native forest are reflected onto the surface of the water—a sort of triangulation that just feels right.

The clean lines of the pool and the concrete paving interact with natural elements, such as native boulders and the arching branches of the coast live oak trees. Steps lead from the pool to a pool bath and other more rustic areas of the property.

LEFT Sunlight filters through oak branches, casting a lattice of shadows on the front entry courtyard and creating a serene and welcoming arrival. A few stone boulders partially buried below grade look as if they have always been there. Swaths of Atlas fescue and Santa Barbara sedge line the pathway, with small stands of *Iris douglasiana* adding a little color when they bloom.

RIGHT Five concrete steps, poured directly next to a large, well-placed boulder, descend from the home's main living area to the pool deck. Grasses resembling the property's native plants soften the hard lines of the concrete walls and home. Atlas fescue and Santa Barbara sedge sway in the breeze and add movement without growing so tall they obstruct views.

LOWER RIGHT The home's kitchen and dining area are additions that open onto a small terrace and steps leading to the pool. Lucas used an understated palette of materials—mainly concrete paving and walls that had a small amount of black color added so they would not feel stark.

UPPER LEFT A quiet seating area is located off the master bedroom, under the branches of an existing oak tree. Simple materials (gravel and grasses) blend with the native forest—it's hard to tell where the landscape ends and the wildland begins.

LEFT Lucas moved the parking area away from the home to create a quiet arrival experience and a sense of refuge.

RIGHT The line of this wall leads to one of the oaks that bookend the pool area, while the stairs point toward the grand vista over the pool.

Before Lucas began work on this property, there were virtually no flat, open areas. Lucas used the pool as a retaining system, creating a much-needed space for entertaining, swimming, and enjoying dramatic views.

CHAPTER TEN

Among the Olives

The lowest branches of the stately olive trees were trimmed to create a spot for an outdoor dining area, kitchen, walkways, and seating. Only three olive trees were removed and relocated to other areas of the garden to create larger spaces within the existing olive grove.

Under a towering canopy of olive trees, which are more than one hundred years old, is a lush landscape with an old-world Mediterranean feel that surrounds a French country–style home. Because Mike Lucas got involved in this project during the early stages of the home's remodel, he had plenty of time to get to know the owners and to explore the olive grove, the native hillside behind the home, and the outlooks that provide views of the valley and local vineyards. "Being tucked into the hillside creates a sense of prospect and refuge, as if you're being protected by the land and the canopy of trees while looking out at the view," says Lucas. The design developed from this concept, with seating areas, terraces, and the pool tucked into the hill, facing the vistas and building on this sense of safety.

Though the homeowners' wish list included a pool, outdoor dining areas, a new entryway, and firepit areas, several early decisions drove the design. "I knew keeping the olive grove intact was important to preserve and protect the character of the property," says Lucas. He also knew there was no way to include a pool at the back of the house, as is traditional, because of the shade of the olive trees and the site's steep terrain. With this in mind, Lucas developed a network of trails that connects the terraces meant for entertaining, arranged at various hillside elevations, with the pool area at the bottom.

Some of the elevation changes are subtle—sometimes only a few steps. "Even a couple of steps is enough to define a space and separate it," says Lucas. Tucking gravel patios and walkways where they fit best into the grade below the olive trees ensured that they wouldn't need to remove any trees and that the tree roots would be protected as much as possible. Lucas placed the dining patio near the home, just a few steps down from the indoor kitchen, to provide easy access through the French doors. A few steps down from the outdoor kitchen is a large built-in seating area with a firepit—another nook among the olives. A couple of the olive trees were relocated to accommodate this addition and to allow more light into the interior of the grove—several were moved to the outer edge of the grid and one to a prominent location by the front gate.

From these main patios, paths extend out in nearly every direction: to the front entry, pool, a more remote bonfire pit (which uses wood instead of gas), lawn, vegetable garden, and outdoor kitchen. "There are many paths aligned with the grid of the trees, but it feels free-flowing and open under the olives as people move up and down from space to space," says Lucas. "One important decision was trimming the lower branches of the olives. This opened up the canopy. I wanted the space under the olives to feel more like a cathedral than a thicket."

In siting the pool, Lucas faced the constraint that the only area on the property that receives full sun throughout the day was in the front yard. "I knew the pool would have to go out front but didn't want it to be the first thing people see when they arrive," remembers Lucas. "The arrival experience needed to be separate."

Lucas decided to install the pool above a new retaining wall and driveway that leads up from the road. This hillside arrival experience features a mix of Mediterranean and native plants that help to set a mood while obscuring the main areas of the landscape.

The driveway leads around the side of the property, winding toward an arrival court with space for several cars to park in front of the house under a few oaks. The main driveway continues around the side of the house to the garage in the back. "Placing the driveway off to the side opened up the garden and reclaimed a large area of the front yard for fun and functional spaces, such as the pool, and mood-setting spaces, such as the walkway to the front door."

The front entryway is accessible through a gap in bay laurel hedges that partially conceal the parked cars. The approach to the front door fully embodies the landscape's old-world French country feel. Lush flowering plants, such as roses, hydrangeas, catmint, and other Mediterranean plants, surround a series of landings. "This first impression captivates people and shows them there's lots to explore and enjoy," says Lucas. The first terrace opens up to views over the pool toward the vineyards. Visitors also get first glimpses of the towering olive grove and paths that weave through the twisted trunks.

"To get the bucolic French style to come through, we needed to use materials that would age gracefully

A pergola covered in lush vines frames the front door. Lucas avoided introducing too much variety and color in nearby plantings, to maintain a feeling of serenity.

The driveway leads homeowners and their guests to a parking area under native oaks. Bay laurel hedges separate this space from the rest of the garden, screening cars from view. A break in the hedges leads to an L-shaped walkway—one arm of the *L* guides visitors to the front door, while the other walkway, at left, leads straight down to the pool.

and add to the scenery," says Lucas. In the entryway approach—and throughout the landscape—lichen and moss make rough-hewn stone steps and pathways appear aged. Vines scamper across a metal pergola over the front door and grow up the side of the house, and the scent of rosemary and lavender waft through the space.

Formal touches balance the more laissez-faire elements. Boxwoods sheared into globes and planted in plinths create a formal entry to the front walkway. Plinths planted with yuccas line the walkway to the front door. "The entry path immerses people in the setting and sets the tone for the entire experience, indoors and out," says Lucas.

From the entryway landings, a path leads to the front door, where a small sitting area to one side offers gorgeous views. "Sitting in a perch with your back surrounded by the safety of the home and the beauty of flower-covered plants just feels good," says Lucas. "I sprinkled intimate seating areas like this throughout the garden."

Heading in another direction on the entry path, the homeowners and their guests can head down a small flight of steps to the olive grove, where a network of gravel paths leads to the pool and other outdoor entertaining spaces. Under the shade of the olive grove next to the kitchen, Lucas created the main hub for entertaining. There's a formal area with stone steps, seating, a firepit, and an outdoor kitchen that all blend with the aesthetics of the house. "These features are all within a few steps from each other and are accessible for daily use," says Lucas. The owners wanted paving that was easy to walk on in these heavily traversed areas, but that meant installing lots of stone. "With all the stone I wanted to add heavy doses of plant life to soften the gravel hardscape and architectural elements," says Lucas.

"The landscape under the olives is quite lush, but it didn't start that way. I wanted to avoid the stone quarry look," remembers Lucas. Under the olive trees there was dense underbrush and shrubbery, but the spaces are now filled with a verdant mix of plants that flower in succession through the seasons. Lucas used boxwoods, rosemary, and other evergreens to continue the design language of French formality throughout the garden and also placed flowering plants, such as hellebores, dogwoods, anemones, irises, roses, sages, hydrangeas, lavenders, and teucriums. "There are layers of concentrated plantings near the home, but as you move farther to the fringes, the planting gets looser and wilder," says Lucas. Native plants, such as deer grass, ceanothus, and oaks, cover much of the landscape near the property's edges—around the lawn and the bonfire pit, at the outer edges of the pool area, and along the driveway retaining wall.

Around the pool, this transition from lush Mediterranean plantings to dry California natives is most visible. The white roses, rosemary, Italian cypress, lavender, and catmint planted on one side of the pool give way to a large swath of natives such as the tawny deer grass on the other side. "I wanted the plants to match the rest of the view when the homeowners are in the pool area," notes Lucas. "Looking toward the house, the plants fit the French country style, but the grasses are more fitting for the view that overlooks the vineyards (and hold the place for a future pool house, now being constructed). They don't distract but provide a simple foreground for the view while we wait."

"The owners of this property were easygoing and open to my ideas," says Lucas. "It empowered me to really go for it and test how far I could go to express the French country style that the home personified." He also notes that the constraints of a project (in this case, the olive grove, the topography of the land, a septic system, and the architectural style of the home) often lead to the best expressions of the landscape's true character. "When designing, you can't put your blinders on and pretend these things aren't there," advises Lucas. "You have to explore, inquire, and respond to the property and the homeowners to discover and then fully express the essence of a garden." ☼

The front walkway and porch overlook the pool and Sonoma Valley. The pool was placed in the front yard because it was the only area that receives full sun during the day (the rest of the property is covered in shade from olive and oak trees). It's raised slightly above the level of the driveway and entry gate (behind the hedges on the left) to preserve the views of the valley over 'Iceberg' roses and other perennials.

1. Guest Parking (Two Areas)
2. Residence
3. Covered Entry
4. Garage/Guesthouse
5. Firepit
6. Olive Grove
7. Kitchen Garden
8. Pool
9. Firepit (Knoll)
10. Lawn (Knoll)

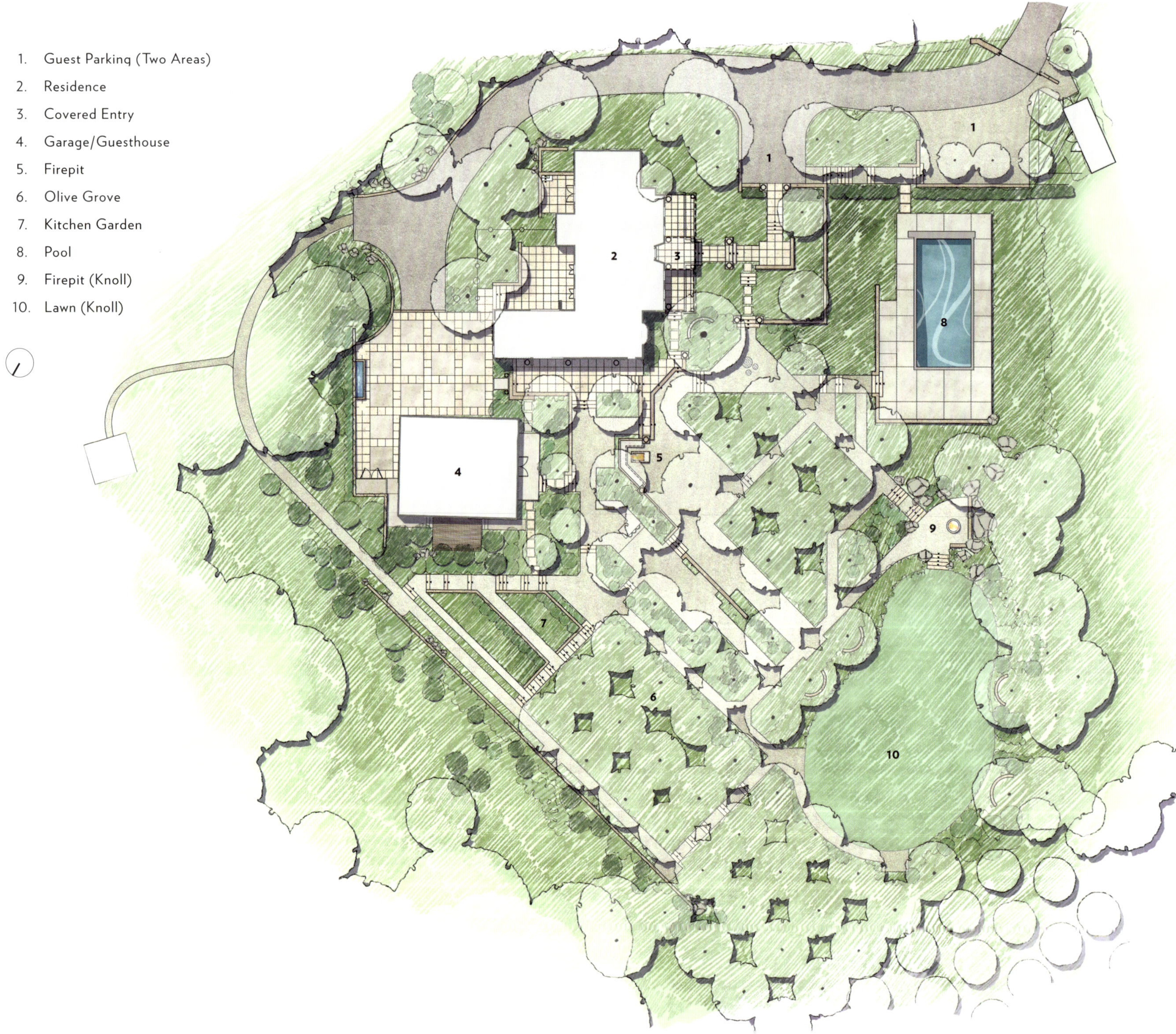

ADVICE FROM LUCAS

Sketching and Developing Important Site Features

The more time you spend on a property, the more you will understand it. Spending time will allow you to identify the best views, the most interesting ways to walk through the land, the microclimates, and so much more.

Sketching is one of my favorite ways to understand a site. I sketch everything from the existing slopes and trees to the spaces I envision walkways winding through. Even details such as where prevailing winds come from make it into my sketches and diagrams. While photographs are helpful in their own way, sketches clarify what's really important. When I get back to the drafting board, my sketches bring to mind vivid memories.

On this project, I walked and embodied the olive grove at least a dozen times, imagining how trails would align with views. I developed a sense of how paths, seating, dining areas, a bocce court, and an outdoor kitchen would fit within the existing grid of olives. Through this process of walking and observing, I gained an appreciation for how the canopy of olive branches created cathedral-like corridors. That experience inspired me to design pathways on which the owners could experience a similar feeling of excitement and wonder when they stroll below these majestic trees.

LEFT A beautiful outcropping of native tufa stone on the side of the knolltop was transformed into a firepit that overlooks the pool. This campfire space, surrounded by oaks, coffeeberry shrubs, and native grasses and perennials, feels worlds away from civilization and even the rest of the garden.

RIGHT A transitional space connects the front of the house to the main entertaining terraces among the olive trees. Lucas designed a more formal layout close to the home, with stone steps and walkways, as well as sheared boxwood hedges and globes. Farther from the house, informal mixes of plants are left to grow naturally.

LOWER RIGHT The entryway to this French country–style home sets the stage for what's to come. Fragrant Mediterranean plants, such as rosemary and lavender, are paired with flowering white roses and hydrangeas. A sandstone capped seat wall creates space for greeting visitors. Yuccas in bowl-shaped planters frame the walkway to the front door.

LEFT Lucas used Sequoia buff stone for this firepit seating area and throughout the garden; the stone develops a patina over time, contributing to the aged country aesthetic. A small sheared boxwood hedge reinforces the architecture and adds softness to this otherwise hardscape-filled space.

RIGHT Lucas located the pool on a terrace created by a retaining wall built into a slope that extends several levels below the front of the house, maintaining both privacy and wonderful views.

LEFT A shady seating area under the pergola provides refuge and protection from which to enjoy views of the valley and vineyards below.

RIGHT The steps from the entry walkway are lined with lush plantings of germander, rosemary, and white roses. Creeping thyme grows between gaps in the sandstone pavers, and bowls containing *Beschorneria* lead the way down toward entertaining spaces under the olive grove.

Mediterranean shrubs, tall Italian cypress, deer grass, and bay laurel hedges separate the pool from the parking area.

CHAPTER ELEVEN

Wine Country Wonderland

The mix of this meadow is simple, with just five varieties of plants. *Pennisetum spathiolatum*, *Muhlenbergia dubia*, *Festuca mairei*, *Gaura lindheimeri*, and *Verbena bonariensis* all have wispy growth habits, so they sway in the breeze, adding continual movement to the garden.

The owner of this expansive five-acre property in Napa Valley wanted to create a landscape that embodies the mood of wine country. "The owner is in the wine industry and has a gracious, warm, and welcoming personality," says Mike Lucas. "She loves having large gatherings, so the design reflects that spirit." The landscape, located in a field with massive valley oak trees, is a celebration of food, wine, culture, friends, and the California lifestyle.

"I wanted the property to have an organic richness and simplicity," says Lucas. "We focused on organizing the space and defining areas for different activities. Otherwise, we used a very light-handed approach to the design." Throughout the front yard and backyard, walls and plants were used to separate spaces, block unsightly views, and create distinct entryways to paths that lead through the landscape. Otherwise, there wasn't much heavy lifting; one of the benefits of developing a flat property is that it requires less grading, retaining, and engineering.

In the front of the house, paths lead to the front door and a side gate. Lucas notes that the side garden gate is used as much as—or more than—the front door, which is somewhat obscured behind landscape walls and large olive trees. The design funnels visitors toward the garden gate and a central path that leads through the backyard and splits off to each of the outdoor spaces. "From this gate, the entire landscape opens up—it's like entering a wonderland that keeps unfolding," says Lucas. The gravel sitting area under the oak at the back of the house, vegetable garden, bocce court, pool, and pool house are all connected to that linear central walkway.

Upon entering through the gate, you can walk straight to the pool and pool house, where large overhangs create ample shade for lounging and an indoor-outdoor kitchen and living room provide spots for relaxing. "Every comfort is easily accessible at the pool and pool house," says Lucas. "These creature comforts were necessary because it's a trek to get back to the house."

Lucas sited the pool and pool house at a distance from the remodeled home, on the sunniest part of the property. As the home is a full-time residence, the owner wanted to create a destination that would be a retreat from daily life. Additionally, large oaks right off the back of the house would have presented several challenges. "The pool couldn't go under those larger trees because they create too much shade and leaf litter," says Lucas. "Plus, disturbing their roots can destroy the trees."

Locating the pool toward the back of the property permitted serene views of the house and the oaks that tower over the land. "You're surrounded by these majestic oaks in nearly every direction," says Lucas. "The trees are massive and have such presence that I based much of the design on their scale and style—it's all proportional to the giant trees." From the wide paths to the sprawling gravel patios, the long bocce court to the wide-open fields of ornamental grasses and perennials, the large dining table to the lengthy landscape walls, the features are oversized so they aren't dwarfed by the trees or the property's length.

Another benefit to siting the pool near the back of the property was that it anchored the landscape and presented opportunities to create destinations throughout the garden and make full use of the property. All these destinations are organized and connected to the central path. Close to the house, a vegetable garden and orchard sit on one side and a shady gravel patio under the oaks on the other. Farther along the path, a long bocce court runs parallel to the walkway on one side, with a large grassy meadow on the other. "The central path links all the elements on the property, acting as a bridge between spaces so you always have a reference point," says Lucas.

"The oak trees also provide context and act as navigational beacons," he continues. "Because the oaks at the back of the house are so large and distinct, you always know where you're at when you lose sight of the home—and there's just one central path that leads back." This simplicity extends to other features throughout the landscape. Paths and patios are covered mostly by gravel or random flagstone and the pool area by sandwashed concrete. Seat walls are constructed of a local quarried stone that blends in with the color palette of the gravel, grasses, and oaks.

This frequently used outdoor dining table is set under the pool house's overhanging canopy. Large doors open to an indoor-outdoor kitchen and an indoor lounge. Pots planted with agaves and lomandra soften the black metal columns of the structure. The agaves' shape, color, and texture add energy and variety to the meadow grasses.

Evening sunlight illuminates the meadow's ornamental grasses. The pool, like the stone seat wall, was proportioned to the size of the oaks. Plants surround the rectangular pool and pool deck on three sides, softening the clean lines.

Lucas kept the varieties of plants to a minimum. Only five types of plants grow throughout much of the front yard and backyard: Atlas fescue (*Festuca mairei*), Lindheimer's beeblossom (*Gaura lindheimeri*), pine muhly (*Muhlenbergia dubia*), slender veldt grass (*Pennisetum spathiolatum*), and tall verbena (*Verbena bonariensis*). These plants form large, rambling meadows bursting with tall grass blooms. The flowering perennials add touches of color in pinks and purples throughout the season. "Atlas fescue is an especially important plant when creating these meadows because it's evergreen," says Lucas. "It holds the space when other grasses need to be cut back in the winter or out of season." Lucas also added a few olive trees in the front yard, with bay laurel hedges to block unsightly views and agaves in strategically placed containers set in a rhythmic pattern along the central walkway. "The agaves also have a bold, sculptural feel that balances and contrasts with all the wispy grasses," says Lucas. "The containers bring some punctuation and order to the free-flowing grasses."

The simplicity of the plant palette and all other architectural features was key to creating the relaxed modern vibe the owner desired. "I didn't do much architecturally in the landscape," says Lucas. "Besides the stone walls and the metal fences, the only other structural feature is the black metal lights we designed that overhang the bocce court." These lights are painted a similar color to the outside of the pool house, so they connect the spaces visually. They allow the owner to play at night—which she does frequently as she enjoys a glass of wine made from grapes harvested from the vineyard out front.

"From the moment people enter this property and begin driving through rows of sauvignon blanc grape vines, there's a feeling of warmth and generosity," says Lucas. "It feels as if you're being welcomed with open arms." Visitors encounter chickens roaming, vegetables overflowing, grass swaying, and oaks providing shade, with native birds frolicking in the branches. "The scenery and environment delight the senses in every way," says Lucas. "It is easy to socialize while also appreciating the bounty of the earth in this Napa landscape." ✵

1. Auto Court
2. Barn
3. Vineyard
4. Main Residence
5. Garage
6. Garden Gate / Side Gate
7. Central Walk
8. Gravel Court
9. Bocce Court
10. Pool House
11. Pool
12. Outdoor Kitchen
13. Kitchen Garden
14. Oak Grove

ADVICE FROM LUCAS

Creating Views

Create views within the garden. Stunning views don't exist on every project. The properties I work on may not overlook native forests, rolling hills, sprawling vineyards, or oak woodlands. But you can create your own visuals: a pool area with views of large oak trees over a meadow-like planting, a seating area under an important tree, or a walkway through soft grasses.

Other than a glimpse of hills to the east, there were no distant views to work with in this garden. We created our own views by using the large existing oaks and by installing architectural elements. Fencing, a pool house, trees, walls, and tall grasses surround the living area in all directions so that the backyard feels like its own world.

Internal views heighten the importance of every element. Plants, materials, and finishes receive more up-close attention when distant views aren't a focal point. From the custom stone walls to the architectural facade of the pool house, the containers along the walkways to the lighting selected for the bocce court, and the rustic metal fencing to soft flowy grasses, I created compelling vignettes and took great care that architectural features and plants in every direction integrate seamlessly with the overall atmosphere.

LEFT The modern-style pool house, designed by Bevan & Associates, has everything the owner needs for entertaining guests or lounging by the pool. An outdoor kitchen, pizza oven, and grill are located next to a large overhang that provides shade for abundant seating.

RIGHT The bocce court, framed by effervescent grasses, runs parallel to the long walkway leading from the home to the pool. Here, looking toward the kitchen garden, Lucas installed overhanging lights so the owner could play in the evenings.

LEFT Containers of *Agave desmettiana* and *Lomandra longifolia* 'Breeze' create a rhythmic pattern along the pathway. They also serve a specific design purpose, adding structure and repetition to the soft wispiness of the grasses.

RIGHT Viewed through the side gate, a corner of the vineyard peeks out from behind the metal fence. The see-through fence creates an open feel and keeps the vineyard in view from most of the backyard. To the right, beyond grasses and the metal fence, are the vegetable garden and orchard.

LEFT The top of the courtyard seat wall is wide and inviting. Since the property's trees are huge, Lucas scaled the architectural landscape features to match. Walls are long, openings are wide, paths are generously sized, and the meadows are expansive.

RIGHT A large pizza oven next to the outdoor kitchen was constructed with board-formed concrete. A black metal pergola provides shade in the grilling and cooking space.

LEFT The driveway leads through the owner's sauvignon blanc vineyards toward the arrival area and a barnlike structure. The oaks, aglow from the sunlight, beckon visitors to explore and enjoy the beauty of the land.

LOWER LEFT Inside the metal side gate, concrete pavers lead to a crushed gravel trail and patio space. A whole world opens up, filled with gorgeous plants, a vegetable garden, an orchard, shaded seating under large oaks, an outdoor kitchen area, and a swimming pool and pool house.

RIGHT The front door to this Napa Valley home is somewhat obscured by deer grasses, stone walls, and large olive trees. The owner, who has guests over frequently, considers both the front entrance and the side gate to be the front doors of her home.

Tough, Multipurpose Plants for Low-Water Landscapes

MIKE LUCAS

This meadow separates the pool area from a gravel patio under an oak. *Pennisetum spathiolatum*, *Gaura lindheimeri*, and *Verbena bonariensis* combine in a wavy sea of plants behind two chairs.

While the architectural features of the garden are what we have become known for at Lucas & Lucas, plants are the icing on the cake. Homeowners develop connections to plants. Trees, shrubs, grasses, ground covers, and flowering perennials powerfully change a garden's visuals, softening the hard edges of built structures. Without plants to obscure walls, to frame views, to adorn the side of a house, to block the wind, and to add fragrance or lushness, the structures we build would look disconnected from the land—stone or concrete monoliths jutting out from bare sites.

Plants personalize a landscape and give it character. Whether I'm preserving existing oaks that frame a view and catch the sunlight just right or adding vibrant flowering perennials among ornamental grasses covering a large hillside, the purpose is the same—to choose plants that fit the character of the site and evoke an appropriate connection to the place.

In the end, I can design all the best walls, hardscapes, walkways, and pools, but plants are what owners expect to see and what our work is judged on. Over the years, I've found a handful of plants that are particularly fitting for the landscapes I design. My favorite plants are tough and adaptable, often capable of surviving in low-water situations and at sites where deer are present. What follows are fourteen of my favorite plants, why I like planting them, and how they are used in landscapes I design. I've narrowed down this list over the years as I've returned to projects I've installed. Year after year, these are the star performers.

Plants can make or break a space. They can be used to connect hardscapes to the land, screen views, catch sunlight, attract wildlife, and add movement, texture, and so much more.

1. *Laurus nobilis*

(BAY LAUREL)

This deep green, upright shrub can also be grown as a tree. Sometimes called Grecian laurel, these evergreens are drought-tolerant Mediterranean natives. When planted as columns, they make a nice, informal screen, background plant, or formal hedge. They thrive under the canopy of other trees or in full sun. I love this plant for its versatility and ability to dance in and out of a wide range of growing conditions. With its lanceolate leaves and contrasting dark green color, it combines well with olive trees. Zones 8–10

2. *Lomandra longifolia* 'Breeze'

(DWARF MAT RUSH)

I've fallen in love with this grasslike evergreen for its chartreuse color and midsize, fountain-shaped growth that always looks good. It's a true workhorse in the garden. It works well in planting designs as a stand-alone clump or in large masses combined with any number of accent plants, such as agaves, verbenas, sages, kangaroo paws, palms, olives, or westringia; it's the perfect plant to use as a unifying filler. It grows to three feet tall and wide in full to partial sun. Zones 8–11

3. *Nepeta* 'Walker's Low'

(CATMINT)

Nepeta is another tough mainstay in many of the gardens I design because of its ability to grow in a range of soil conditions and its drought tolerance. Its lavender-colored blooms are abundant in May and June. With a hard cutback in summer, it puts on another show later in the season. From a style perspective, it pairs well with salvias and fits into formal landscapes and also in Mediterranean or naturalistic gardens along paths. Nepeta grows best in full sun, and its size reaches two feet tall by two to three feet wide. Zones 4–8

4. *Olea europaea*

(COMMON OLIVE)

I love planting mature olive trees because they deliver the best bang for the buck, providing instant presence and stature. Olive trees fit in with many California landscapes and are often found naturalized in California hillsides. Nothing beats transplanted olive trees from orchards in the Central Valley to give a garden a sense of instant maturity or antiquity. Trees reach fifteen to twenty feet tall and blend well with oak woodlands, where their branching patterns mirror those of the coast live oak. Trees that are over one hundred years old are stout in their structure. They can grow to more than twenty-five feet tall and are perfect for lending an old-world feel. Their textured and glaucous leaves, sculptural trunks, and drought tolerance make them a winner in almost every situation. Zones 8–11

5. *Olea* 'Little Ollie'

(DWARF OLIVE)

Another of my go-to varieties is *Olea* 'Little Ollie'. It has a wonderful texture, round structure, and nice gray color. It's also deer and drought tolerant. This hardworking, sun-loving evergreen shrub, which grows to four to six feet tall and wide, is a mainstay in many of my designs because it fits well with a number of plant palettes—from formal to naturalistic. Zones 8–11

6. *Pennisetum spathiolatum*

(SLENDER VELDT GRASS)

This showy bunchgrass forms tight mounds with flower spikes that grow three feet above the foliage. Tan-brown flowers emerge in April and last until fall, displaying a color that mixes well with the grassy hillsides abundant throughout California. Slender veldt grass works well in large masses or small drifts in naturalized meadows, where it can flow into the shade of trees (though it may bloom less profusely there). I use it for a dramatic effect in large spaces, where backlighting and breezes accentuate the tall stalks that sway to and fro—it can really put on a show. I particularly like using *Pennisetum spathiolatum* adjacent to annual grass fields, as it blends in nicely and blurs the line between the maintained and unmaintained landscapes. I also often plant it along walkways with verbena, gaura, and other low-growing plants. In the winter, this grass responds quickly to cutback and pushes out neat new growth almost immediately, so it shines while other plants sleep. Zones 8–11

7. *Heuchera maxima*

(ISLAND ALUM ROOT)

Island alum root is native to the Channel Islands of California. It forms clumps of evergreen foliage and grows one foot tall by two feet wide. Pinkish-white flowers bloom in spring, rising several feet above the foliage; I look forward to this spring show every year. This heuchera grows best in cool sun or partial shade, so it's the perfect plant to place under oaks and in the shadier parts of a meadow. Massing it along trails provides quite a punch of color, and its foliage maintains this color through the dry season and into winter. Zones 8–10

8. *Frangula californica*

(CALIFORNIA COFFEEBERRY)

This is one of my favorite California natives because it has a neat, clean appearance. California coffeeberry is an evergreen shrub with deep green, glossy leaves that are perfect as a dense screen or windbreak. It can be used to break up walls with some greenery or to hide unsightly elements, such as pool equipment or utilities. It grows to about eight feet tall and wide in sun or shade. It needs little water once established and works well planted around oaks. Zones 7–11

9. *Rubus pentalobus*

(CREEPING RASPBERRY)

Creeping raspberry is an evergreen ground cover that looks beautiful spilling over walls, growing under oaks and around boulders, and covering large areas of dry ground on slopes. It thrives in rocky soils with little water once established. I love how quickly this plant fills in across an area, eventually growing to around one foot tall by four feet wide. It brings a natural yet tidy feeling to areas of the garden that are difficult to plant, such as spaces under oaks. This ground cover is tough and doesn't require much maintenance. Plant it in partial shade. Zones 5–9

10. *Salvia nemorosa* 'Caradonna'

(MEADOW SAGE)

This herbaceous perennial blooms from spring through summer. Planted in meadows among ornamental grasses or along border trails, this sage has upright, dark purple flower spikes that can reach two feet tall. The flower's rich color makes it a nice addition along walkways. I love it for its spring show; if you cut it in early summer after the first bloom period, it comes back again in late summer. This sage grows best in full sun, in dry to medium soil. Zones 8–11

11. *Sesleria autumnalis* 'Greenlee'

(GREENLEE MOOR GRASS)

Greenlee moor grass is a low-growing, one-foot-tall-by-two-foot-wide grass. This versatile plant can act as a unifying ground cover and a substitute for a lawn. It needs a little more water than some ornamental grasses, but its lush evergreen foliage, tawny flowers, and low stature make it a good choice as a base plant for a meadow or a covering for large areas along paths. It grows best in full sun but can handle some filtered shade. This grass does well along trails and in spots where you need to transition to larger perennials. Zones 5–9

12. *Symphyotrichum oblongifolium* 'Fanny'

(AROMATIC ASTER)

This aster has dark purple, daisylike flowers that bloom in October and continue until frost. I find it gorgeous planted in and along the edges of meadows, where it can be enjoyed from walkways, patios, and seating areas. It grows to three to four feet tall and wide in full sun to partial shade. Zones 5–9

13. *Teucrium chamaedrys*

(WALL GERMANDER)

This low-growing evergreen shrub is a true favorite of mine and works well in many situations. It provides neat, fine-textured, deep green foliage—a look that goes well with formal plantings when planted as a small hedge or, for more naturalistic settings, planted in sprawling swaths. If not cut back regularly, it will grow to about one foot tall and three feet wide. Its mauve flowers that bloom in midsummer are a nice bonus. Grow in full sun. Zones 5–9

14. *Verbena bonariensis*

(TALL VERBENA)

This is a wonderfully tough perennial. Its flowers bloom in my favorite shade of purple; I find it soothing in the midsummer heat. Since the flower stalks tower above many ornamental grasses and other perennials, they can be easily enjoyed. It has a beautiful silhouette that's most visible when backlit by the sun. Because *Verbena bonariensis* has an airy growth habit, it sways in even a gentle breeze. While it's a relatively short-lived perennial, it can seed nicely around the garden, creating a spontaneous feel and a pop of purple color in stylized meadow plantings. Grow in full sun to partial shade. Zones 7–11

Afterword

JENNIFER LUCAS

Landscapes designed by Mike Lucas almost escape description. It's as if he has the ability to bring forth the soul of a place. With little more than descriptions of preferred aesthetics and a wish list, Mike is able to deeply understand our clients. His intuitive approach is complemented by his commitment to fully understand the land, site conditions, and desired uses—Mike is known to camp out on project sites before design begins, to feel the land and to begin to understand potential design solutions. This leads to finished projects that are experiential, beautiful, unique, calming, and connected to their site, to their home, and to their users.

He has been designing this way since we first met in graduate school at Cal Poly Pomona in 1998. I was in the architecture program, and Mike was in the landscape architecture program. During the first quarter, the two programs were combined, which made perfect sense to us then and still guides our practice today. We know that a landscape is not at its best when designed in isolation from the architecture, and architecture that does not come from its site is not grounded or even easy to construct. At our firm, architecture and landscape are inseparable, and we create integrated, holistic spaces that relate to their place and provide respite for their users. Our design solutions are right at home here in Northern California, where we are fortunate to be able to live, work, and play outdoors year round.

Our enthusiasm for built forms that are beautiful and integrated with the land, seamless movement between indoors and out, framing and highlighting near and distant views, and living life outside in the garden informs each of our projects. The highly inhabitable spaces featured in this book showcase the results of a thoughtful approach to developing a landscape where home, garden, and site are woven together into a seamless whole. These gardens are a pleasure to experience, spark curiosity and joy, and are restful for the soul. We hope you enjoy these landscapes, lessons, and plantings and that they help inform decisions as you create architectural spaces of your own.

Acknowledgments

To my wonderful wife, Nicole, for all her years of support and encouragement.

To my kids, Maren, Emmett, and Tessa. (The book is done. Now daddy can play.)

To Jim Peterson, who gave me the opportunity to be editor of *Garden Design* magazine and taught me how to manage projects.

To Justine Guidry, who is my go-to editor and comma queen.

To the late Bill Marken, who taught me how to look at landscapes and write about the world's best gardens.

—Thad Orr

A warm thanks goes out to all those who have helped make this project possible. First of all, to our clients, who have entrusted us with the privilege of serving them, thanks for believing in the process. We always have your best interests at heart. To my first boss, Mark Bowers, for taking me on when I was without practical skills and showing me the rules of the road. And to Gary Ratway, who showed me how to break those rules and dream big. Both lessons have served me well. A special thanks also to those I have worked with along the way, both in our office and out in the field—those who help bring our plans to life. Without them, there would be no gardens. My hats off to Thad Orr for suggesting this monograph; he saw the possibility and had the gumption to see it all through. Without him this book would not be here. To Princeton Architectural Press for believing in our work and to Marion Brenner and Caitlin Atkinson for bringing it to life with their incredible eyes, the light and space that they see, and the truthful story that they tell with their photography. Thank you.

Most important, I want to thank my wife and life partner, Jennifer Lucas. She is my one. Thank you for your belief in me from the start, for seeing what is possible and providing me with an all-important grounding force in my life. You are my foundation, my mirror, my true love. You have sacrificed so much. It has not gone unnoticed, and I thank you for our greatest creation of all: our children, Owen and Flora. With them we have built and share our safe nest on this wonderful planet. It's truly a better place with you here by my side.

—Mike Lucas

Appendix I: Credits

Chapter 2: Knolltop Retreat
DESIGN TEAM: Mike Lucas and Emily Podolak
ARCHITECT: Nina Nazarov Architect
GENERAL CONTRACTOR: Total Concepts
LANDSCAPE CONTRACTOR: Oxbow Pool and Landscape Contractors
SWIMMING POOL CONTRACTOR: Swan Pools
PHOTOGRAPHER: Caitlin Atkinson

Chapter 3: Private Sanctuary
DESIGN TEAM: Mike Lucas and Emily Podolak
ARCHITECT: Bevan + Associates
GENERAL CONTRACTOR: Benchmark Construction
LANDSCAPE CONTRACTOR: Ray's Landscape
PHOTOGRAPHER: Marion Brenner Photography

Chapter 4: Hillside Modern
DESIGN TEAM: Mike Lucas and David Fowler
ARCHITECT: Bevan + Associates
GENERAL CONTRACTOR: Benchmark Construction
LANDSCAPE CONTRACTOR: Siteworks Landscape
POOL BUILDER: Wine Country Pools
PHOTOGRAPHER: Marion Brenner Photography

Chapter 5: Old Meets New
DESIGN TEAM: Mike Lucas and Emily Podolak
BUILDING DESIGNER: Julian Cohen / ARC Design
GENERAL CONTRACTOR: Monsoon Construction
LANDSCAPE CONTRACTOR: Dexter Estate Landscapes
POOL BUILDER: Stan Johnson Pools
PHOTOGRAPHER: Marion Brenner Photography

Chapter 6: Treetop Hideaway
DESIGN TEAM: Mike Lucas, Emily Podolak, Greg Dalton, Jessica Norling, and Tori Johnson
ARCHITECT: Cahill Studio
GENERAL CONTRACTOR: Robinwood Construction and Landscape
LANDSCAPE CONTRACTOR: Robinwood Construction and Landscape
POOL BUILDER: Stan Johnson Pools
PHOTOGRAPHER: Caitlin Atkinson

Chapter 7: Prairie Leisure
DESIGN TEAM: Mike Lucas, Tori Johnson, and Greg Dalton
BUILDING DESIGN: Patrick Printy Design
GENERAL CONTRACTOR: JG Milne Builders
LANDSCAPE CONTRACTOR: Oxbow Pool and Landscape Contractors
SWIMMING POOL CONTRACTOR: Oxbow Pool and Landscape Contractors
PHOTOGRAPHER: Caitlin Atkinson

Chapter 8: Traditional Estate
DESIGN TEAM: Mike Lucas and Emily Podolak
ARCHITECT: Bevan + Associates
GENERAL CONTRACTOR: Robinwood Construction and Landscape
LANDSCAPE CONTRACTOR: Robinwood Construction and Landscape
POOL BUILDER: Stan Johnson Pools
PHOTOGRAPHER: Marion Brenner Photography; image on pages 148–49 by Cesar Rubio

Chapter 9: Native Refresh
DESIGN TEAM: Mike Lucas, Michael Bliss, and Greg Dalton
ARCHITECT: Strening Architects
GENERAL CONTRACTOR: Holly & Associates
LANDSCAPE CONTRACTOR: Oxbow Pool and Landscape Contractors
POOL CONTRACTOR: Oxbow Pool and Landscape Contractors
PHOTOGRAPHER: Caitlin Atkinson

Chapter 10: Among the Olives
DESIGN TEAM: Mike Lucas, Emily Podolak, Greg Dalton, and Tori Johnson
ARCHITECT: Bevan + Associates
DESIGNER: Patrick Printy Design
GENERAL CONTRACTOR: Gracie Construction
LANDSCAPE CONTRACTORS: Dexter Estate Landscapes, Siteworks Landscape, and Oxbow Pool and Landscape Contractors
POOL CONTRACTOR: Oxbow Pool and Landscape Contractors
PHOTOGRAPHER: Caitlin Atkinson

Chapter 11: Wine Country Wonderland
DESIGN TEAM: Mike Lucas and Tori Johnson
ARCHITECT: Bevan + Associates
POOL HOUSE BUILDER: Elder Lane Construction
LANDSCAPE BUILDER: Oxbow Pool and Landscape Contractors
POOL BUILDER: Paradise Pools
PHOTOGRAPHER: Caitlin Atkinson

All plan illustrations by Eric Arneson, topophyla.com

Appendix II: Plant Lists by Project

Chapter 2: Knolltop Retreat

TREES

Arbutus 'Marina'
Arbutus unedo
Cercis canadensis
Quercus agrifolia
Quercus douglasii
Quercus lobata

SHRUBS AND PERENNIALS

Achillea millefolium
Achillea millefolium 'Terra Cotta'
Agastache cana
Arctostaphylos densiflora 'Howard McMinn'
Correa 'Ivory Bells'
Corylus cornuta ssp. *californica*
Epilobium canum
Erigeron karvinskianus
Garrya elliptica 'Evie'
Gaura lindheimeri 'Whirling Butterflies'
Heuchera maxima
Iris douglasiana
Lobelia tupa
Lonicera nitida
Pennisetum spathiolatum
Ribes sanguineum
Ribes speciosum
Rubus pentalobus
Salvia greggii 'Furman's Red'
Salvia nemorosa 'Caradonna'
Salvia spathacea
Sarcococca hookeriana var. *humilis*
Sesleria autumnalis
Symphytotrichum ericoides 'Lovely'
Symphytotrichum oblongifolium 'Fanny'
Verbena bonariensis

Chapter 3: Private Sanctuary

COURTYARD

Agapanthus sp.
Chionanthus retusus
Citrus limon 'Meyer Improved'
Osmanthus fragrans
Parthenocissus tricuspidata
Philadelphus 'Belle Etoile'
Pittosporum tobira
Pittosporum tobira 'Wheeler's Dwarf'
Pyrus 'New Bradford'
Rosa (std. Apricot / Salmon color flower)
Sarcococca ruscifolia
Trachelospermum jasminoides

FRONT YARD

Agave weberi
Alyogyne huegelii
Anigozanthos 'Orange Cross'
Callistemon citrinus
Elaeagnus pungens
Eucalyptus cinerea
Eucalyptus cladocalyx
Geijera parviflora
Hardenbergia violacea
Kniphofia 'Malibu Yellow'
Lomandra longifolia 'Breeze'
Lonicera japonica
Olea 'Little Ollie'
Pittosporum eugenioides
Salvia 'Mesa Azure'
Verbena bonariensis
Westringia fruticosa

BACKYARD

Quercus lobata

Chapter 4: Hillside Modern

TREES

Acer circinatum
Acer palmatum
Quercus douglasii

NATIVE SHADE

Festuca californica
Frangula californica
Frangula californica 'Eve Case'
Philadelphus lewisii

SHRUBS

Arctostaphylos (spp.)
Baccharis pilularis
Choisya ternata
Cotoneaster dammeri 'Coral Beauty'
Myrtus communis
Romneya coulteri
Spiraea japonica 'Anthony Waterer'
Vitex angus-castus 'Montrose Purple'

GRASSES

Miscanthus sinensis
Muhlenbergia rigens
Pennisetum orientale
Sesleria 'Greenlee'

GROUND COVER

Thymus 'Victor Reiter'

VINES

Hydrangea anomala ssp. *petiolaris*
Vitis californica 'Roger's Red"

PERENNIALS

Achillea 'Walther Funcke'
Agastache 'Blue Fortune'
Lavandula sp.
Origanum laevigatum 'Hopleys'
Penstemon 'Midnight'
Perovskia atriplicifolia 'Blue Spire'
Phlomis tuberosa 'Amazone'
Salvia nemorosa 'Caradonna'
Teucrium chamaedrys

Chapter 5: Old Meets New

TREES

Cercidiphyllum japonicum
Cornus 'Eddie's White Wonder'
Olea europaea
Pistacia chinensis
Platanus sp.

SHRUBS

Buxus sempervirens 'Suffruticosa'
Cornus sericea 'Flaviramea'
Helianthus 'Lemon Queen'
Mahonia aquifolium
Myrtus communis
Myrtus communis 'Compacta'
Philadelphus lewisii
Ribes aureum
Ribes sanguineum
Sambucus mexicana
Sarcococca hookeriana var. *humilis*
Viburnum plicatum

GRASSES
Buchloe dactyloides 'UC Verde'
Festuca californica
Muhlenbergia lindheimeri
Muhlenbergia rigens
Molinia 'Skyracer'
Sesleria 'Greenlee'

GROUND COVERS AND VINES
Aristolochia tomentosa
Asarum canadense
Galium odoratum
Parthenocissus tricuspidata
Ribes viburnifolium
Rosa sp.
Rubus pentalobus
Trachelospermum jasminoides

PERENNIALS
Aster x *frikartii* 'Monch'
Astrantia major 'Canneman'
Brunnera macrophylla
Eutrochium purpureum
Geranium 'Biokovo'
Helianthus angustifolius
Heuchera maxima
Iris douglasiana
Iris germanica
Monarda 'Violet Queen'
Penstemon 'Midnight'
Persicaria amplexicaulis 'Taurus'
Phlomis tuberosa
Romneya coulteri
Thalictrum 'Elin'
Verbascum 'Pink Domino'

Chapter 6: Treetop Hideaway

TREES
Ginkgo biloba
Nyssa sylvatica
Olea europaea
Pistacia chinensis
Quercus rubra
Vitex agnus-castus

SCREENING
Frangula californica 'Eve Case'
Frangula californica 'Mound San Bruno'
Garrya elliptica 'Evie'
Heteromeles arbutifolia
Laurus nobilis

ENTRY MEADOW
Artemisia 'Powis Castle'
Euphorbia mellifera
Iris germanica
Lavandula stoechas
Olea 'Little Ollie'
Phlomis russeliana
Rosa banksiae
Sanguisorba officinalis
Sesleria autumnalis
Sisyrinchium striatum

FRONT COURTYARD
Alchemilla mollis
Arthropodium cirratum
Azara microphylla
Carex divulsa
Cyperus papyrus
Cyperus percamenthus
Fatsia japonica
Lobelia x *speciosa* 'Fan Scarlet'
Lonicera nitida 'Baggesen's Gold'
Mentha requienii
Phlomis russeliana
Physocarpus opulifolius 'Diabolo'
Polystichum munitum
Salvia uliginosa
Sarcococca hookeriana var. *humilis*
Sesleria autumnalis
Sisyrinchium striatum
Woodwardia fimbriata

BACK TERRACE
Arthropodium cirratum
Carex divulsa
Iris douglasiana
Lonicera nitida 'Baggesen's Gold'
Olea 'Little Ollie'
Rosa 'Noare' (Red Flower Carpet)
Rosmarinus 'Collingwood Ingram'
Sesleria autumnalis

PERIMETER MEADOW
Cistus x *hybridus*
Epilobium 'Everett's Choice'
Leucophyllum laevigatum 'Summer Snow'
Muhlenbergia dubia
Pennisetum spathiolatum
Perovskia atriplicifolia
Rosmarinus 'Collingwood Ingram'
Salvia 'Mystic Spires Blue'

Chapter 7: Prairie Leisure

TREES
Chionanthus retusus
Fraxinus americana 'Autumn Purple'
Lagerstroemia 'Natchez'
Olea europaea
Quercus agrifolia
Quercus lobata

ENTRY MEADOW
Frangula californica 'Eve Case'
Leymus condensatus 'Canyon Prince'
Liatris spicata
Lupinus albifrons
Muhlenbergia dubia
Muhlenbergia lindheimeri
Olea 'Little Ollie'
Perovskia atriplicifolia 'Blue Spire'
Rosa rugosa 'Alba'
Rosa spp.
Sporobolus heterolepis
Vitex agnus-castus

DINING COURTYARD
Agapanthus 'Peter Pan'
Buxus sempervirens
Hydrangea paniculata 'Baby Lace'
Rosa 'Mermaid'
Sarcococca ruscifolia

PATH BY BOCCE COURT
Abelia x *grandiflora*
Lavandula angustifolia
Lavandula x *ginginsii*
Nepeta 'Walker's Low'
Olea 'Little Ollie'
Plecostachys serpyllifolia
Rosa 'Sally Holmes'
Salvia nemorosa 'Caradonna'
Viburnum plicatum

PERENNIAL MEADOW

Abelia x grandiflora
Achillea filipendulina
Cistus x hybridus
Frangula californica
Helianthus maximiliani
Miscanthus transmorrisonensis
Muhlenbergia dubia
Pennisetum spathiolatum
Perovskia atriplicifolia 'Blue Spire'
Phlomis russeliana
Salvia azurea
Sesleria autumnalis
Sporobolus heterolepis
Verbascum bombyciferum 'Arctic Summer'

LAWN

'Bolero' fescue sod

Chapter 8: Traditional Estate

TREES

Citrus aurantifolia
Citrus limon
Citrus x meyeri
Cotinus 'Grace'
Ficus carica 'Mission'
Fortunella japonica
Lagerstroemia indica 'Potomac'
Magnolia 'Little Gem'
Platanus x acerifolia 'Columbia'
Prunus armeniaca
Prunus domestica
Punica granatum
Zelkova serrata 'Green Vase'

SHRUBS

Agapanthus 'Midknight Blue'
Agapanthus 'Peter Pan'
Buddleja davidii 'Dark Knight'
Buxus sempervirens
Buxus sempervirens 'Suffruticosa'
Cistus 'Blanche'
Cistus 'Jessamy Beauty'
Cistus 'Snowfire'
Cistus 'Snow White'
Daphne x burkwoodii 'Carol Mackie'
Euphorbia chariacias ssp. *wulfenii*
Hydrangea macrophylla 'Nigra'
Ilex sp.
Lavatera 'Barnsley'
Lavendula x intermedia 'Grosso'
Melianthus major
Myrtus communis
Myrtus communis 'Compacta'
Olea 'Little Ollie'
Phormium 'Maori Queen'
Pittosporum 'Golf Ball'
Pittosporum tenuifolium
Rhododendron 'Horizon Monarch'
Rosmarinus officinalis 'Blue Spires'
Sambucus nigra 'Gerda'
Taxus 'H.M. Eddie'
Teucrium chamaedrys

GRASSES

Calamagrostis acutiflora 'Karl Foerster'
Festuca rubra
Miscanthus sinensis 'Graziella'
Miscanthus sinensis 'Morning Light'
Pennisetum orientale
Sesleria 'Greenlee'
Stipa gigantea

VINES

Rosa 'Cecile Brunner'
Rosa 'Iceberg'
Vitis vinifera

PERENNIALS

Achillea 'Inca Gold'
Achillea 'Marmalade'
Achillea 'Paprika'
Agastache 'After Eight'
Agastache 'Blue Fortune'
Agastache 'Firebird'
Agastache 'Painted Lady'
Agastache 'Tangerine Dreams'
Ajuga 'Chocolate Chip'
Allium 'Gladiator'
Anemone hupehensis
Aster x frikartii 'Monch'
Aster laevis 'Bluebird'
Astrantia major 'Buckland'
Astrantia major 'Claret'
Astrantia 'Moulin Rouge'
Bulbinella floribunda
Crocosmia 'Emily McKenzie'
Cynara scolymus
Cyperus prolifer
Dymondia margaretae
Echinacea purpurea
Erigeron glaucus
Eryngium 'Big Blue'
Eucomis comosa 'Sparkling Burgundy'
Gaura lindheimeri 'Summer Breeze'
Geranium 'Johnson's Blue'
Geranium 'Mavis Simpson'
Helianthus tuberosus
Helianthus 'Velvet Queen'
Heuchera maxima
Iris 'White Swirl'
Kniphofia 'Bees' Sunset'
Kniphofia 'Gladness'
Kniphofia 'Malibu Yellow'
Nepeta 'Six Hills Giant'
Nepeta 'Walker's Low'
Penstemon 'Blackbird'
Penstemon 'Blue Midnight'
Persicaria amplexicaulis 'Firetail'
Phlomis russeliana
Phlomis fruticosa
Pimpinella anisum
Salvia 'El Cielo Blue'
Salvia guaranitica 'Black and Blue'
Salvia nemorosa 'Caradonna'
Sedum telephium 'Karfunkelstein'
Sedum telephium 'Red Cauli'
Stachys byzantina
Stachys coccinea
Thalictrum delavayi 'Hewitt's Double'
Thymus serpyllum 'Minor'
Verbascum chaixii 'Wedding Candles'
Verbascum 'Clementine'
Verbena bonariensis

Chapter 9: Native Refresh

Achillea millefolium 'Salmon Beauty'
Arctostaphylos densiflora 'Harmony'
Carex barbarae
Cercis occidentalis
Festuca mairei
Frangula californica
Iris douglasiana
Polystichum munitum
Rubus pentalobus
Symphyotrichum chilense
Woodwardia fimbriata

Chapter 10: Among the Olives

TREES
Acer palmatum
Aesculus californica
Cornus capitata
Cornus 'Eddie's White Wonder'
Quercus agrifolia

SHRUBS AND PERENNIALS
Agapanthus 'Peter Pan'
Anemone x hybrida 'Honorine Jobert'
Anemone 'Wild Swan'
Arctostaphylos 'Emerald Carpet'
Arthropodium cirratum
Aster x *frikartii* 'Monch'
Buxus sempervirens
Camellia sasanqua 'Setsugekka'
Ceanothus griseus var. *horizontalis*
Correa 'Ivory Bells'
Cotoneaster dommeri 'Coral Beauty'
Daphne x *transatlantica* 'Eternal Fragrance'
Euphorbia amygdaloides ssp. *robbiae*
Euphorbia characias ssp. *wulfenii*
Frangula californica 'Eve Case'
Helichrysum petiolare
Helleborus foetidus
Helleborus x *hybridus*
Hydrangea paniculata 'Limelight'
Laurus nobilis
Lavandula x *intermedia* 'Grosso'
Morella californica
Muhlenbergia dubia
Nepeta 'Walker's Low'
Olea 'Little Ollie'
Philadelphus 'Belle Etoile'
Rosa 'Iceberg'
Rosa 'Mermaid'
Rosmarinus officinalis 'Collingwood Ingram'
Rosmarinus officinalis 'Prostratus'
Salvia 'Bee's Bliss'
Salvia forsskaolii
Salvia nemorosa
Salvia sonomensis
Teucrium chamaedrys
Trachelospermum jasminoides
Viburnum davidii

OLIVE UNDERSTORY
Anemone 'Wild Swan'
Digitalis sp.
Ceanothus griseus var. *horizontalis*
Choisya 'Aztec Pearl'
Correa 'Ivory Bells'
Euphorbia amygdaloides ssp. *robbiae*
Euphorbia characias ssp. *wulfenii*
Frangula californica 'Eve Case'
Helichrysum petiolare
Helleborus foetidus
Helleborus x *hybridus*
Laurus nobilis
Nepeta 'Walker's Low'
Teucrium chamaedrys
Viburnum davidii

NATURAL HILLSIDE
Frangula californica 'Eve Case'
Heteromeles arbutifolia
Muhlenbergia rigens
Salvia 'Bee's Bliss'
Vitis 'Roger's Red'

Chapter 11: Wine Country Wonderland

TREES
Olea europaea
Quercus lobata

SHRUBS
Laurus nobilis

GRASSES AND PERENNIALS
Agave americana 'Variegata'
Agave weberi
Festuca mairei
Gaura lindheimeri
Muhlenbergia dubia
Pennisetum spathiolatum
Verbena bonariensis

POOL HOUSE
Lomandra longifolia 'Breeze'
Olea 'Little Ollie'
Opuntia ellisiana
Pandorea jasminoides 'Alba'

ENTRY GATE
Olea 'Little Ollie'
Pennisetum spathiolatum

CONTAINERS
Agave desmettiana
Aspidistra elatior
Lomandra longifolia 'Breeze'
Muhlenbergia dumosa
Philodendron 'Winterbourn' (Xanadu)

About the Authors

Thad Orr is a writer and editor who enjoys finding and sharing stories about landscape architecture, gardens, and outdoor design. He is the former editor of *Garden Design* magazine and has produced and written stories for publications including *Better Homes & Gardens*, *Sunset*, *Country Gardens*, and *House & Garden*.

Thad enjoys visiting gardens to experience great landscape architecture, meeting the designers who create the gardens, and learning from the homeowners who live in them every day. His favorite stories are captured in the videos he produced with Lucas & Lucas, where viewers can see the gardens and plants as well as hear the designer discuss decisions that were made in the process of creating beautiful landscapes.

Thad currently lives in Redlands, California, with his wife and three kids.

Mike Lucas earned an undergraduate degree in sociology from the University of California, Santa Barbara, and a graduate degree in landscape architecture from Cal Poly Pomona. A Midwestern native transplanted to San Luis Obispo at an early age, Mike grew up surfing and camping up and down California's Central Coast. Mike's personal, soulful approach animates his life and work; his love of sculpting, gardening, family, and wilderness makes landscape architecture a natural fit with his personality.

Jennifer Lucas holds an undergraduate degree in business from Boston University and a graduate degree in architecture from Cal Poly Pomona. She has worked on Wall Street and has lived everywhere from large cities to a cabin in the woods. Jennifer is deeply committed to integrity, client service, and creative design solutions. She brings her urban sophistication and true love of the land, nature, people, and art to inspire refined living solutions in our rural landscape.

Published by
Princeton Architectural Press
202 Warren Street
Hudson, New York 12534
www.papress.com

Printed and bound in China
25 24 23 22 4 3 2 1 First edition

Designer: Paula Baver
Cover Design: Paul Wagner

Library of Congress Cataloging-in-Publication Data
Names: Orr, Thad, author.
Title: Architectural gardens : inside the landscapes of Lucas & Lucas / Thad Orr and Mike Lucas ; photography by Caitlin Atkinson and Marion Brenner.
Description: Hudson, New York : Princeton Architectural Press, [2022] | Summary: "A lushly illustrated portfolio of the gardens of California landscape architecture firm Lucas & Lucas—each project a lesson in the alchemy of garden design"— Provided by publisher.
Identifiers: LCCN 2021029580 | ISBN 9781616899646 (hardcover)
Subjects: LCSH: Landscaping industry—California. | Landscape architecture—California. | Landscape design—California.
Classification: LCC SB472.535.C2 O77 2022 | DDC 712.09794—dc23
LC record available at https://lccn.loc.gov/2021029580